Launching IFMBE into the 2
50 Years and Counting

Herbert Voigt · Ratko Magjarević
Editors

Launching IFMBE into the 21st Century: 50 Years and Counting

Springer

Editors
Prof. Herbert Voigt
Dept. of Biomedical Engineering
Boston University
Boston
USA

Prof. Ratko Magjarević
Faculty of Electrical Engineering
and Computing
University of Zagreb
Zagreb
Croatia

ISBN 978-3-662-52208-0
ISBN 978-3-642-30160-5 (electronic)
DOI 10.1007/978-3-642-30160-5
Springer Heidelberg New York Dordrecht London

Softcover reprint of the hardcover 1st edition 2014

Printed on acid-free paper

Springer is part of Springer Science+Business Media (www.springer.com)

Welcome Address

Welcome to the 2012 World Congress on Medical Physics and Biomedical Engineering from the International Federation for Medical and Biological Engineering (IFMBE). I have looked forward to this event for almost six years, ever since the 2006 World Congress in Seoul, Korea. This will be a very exciting World Congress. It is being held in Beijing, China and has been organized by a wonderful team of cooperative people, who have striven to make this the best Congress it can be. I am thrilled that the IFMBE has contributed many papers, sessions, workshops and special lectures. Highlights of the Congress I am especially looking forward to, are - the **Special Public Lectures** given by the IFMBE Award winners and - the **talks being given by the Biomedical Engineering students** who worked on their Senior Projects involving the development/deployment of low-cost medical devices for use in resource-poor nations. I am very grateful to the **Biomedical Engineering Society, IEEE/EMBS, and *Engineering World Health*** for providing funds, together with IFMBE, to provide travel support to the speakers in this Special Session.

There is much to see and hear, here in Beijing. I hope that your time is well spent learning new knowledge, forming new hypotheses, planning new experiments, developing new health-care delivery techniques, forming and renewing friendships. As a Non-Governmental Organization for the World Health Organization and the United Nations, the IFMBE is uniquely positioned to influence the delivery of health care to the world through Biomedical and Clinical Engineering. As a totally voluntary organization, I challenge you to find a way to help the IFMBE achieve its mission, which is, "to encourage, support, represent and unify the world-wide Medical and Biological Engineering community in order to promote health and quality of life through advancement of research, development, application and management of technology."

Have a great conference!

Herbert Voigt, President, IFMBE

Contents

Part I: Greetings

Part II: Events

Part III: AWARDS

Part IV: Honours – Academy and Honorary Life Members

Part V: Governing Bodies – Officers and AC Members

Part VI: Member Societies

PART I: Greetings

The International Federation for Medical and Biological Engineering (IFMBE) is primarily a federation of national medical and transnational societies. These professional organizations represent interests in and biological engineering. The IFMBE is also a Non-Governmental Organization (NGO) for the United Nations and the World Health Organization (WHO), where we are uniquely positioned to influence the delivery of health care to the world through Biomedical and Clinical Engineering.

The IFMBE's objectives are scientific and technological as well as educational and literary. Within the field of medical, biological and clinical engineering IFMBE's aims are to encourage research and application of knowledge, and to disseminate information and promote collaboration. The ways in which we disseminate information include: organizing World Congresses and Regional Conferences, publishing our flagship journal *Medical & Biological Engineering & Computing* (MBEC), our web-based newsletter – *IFMBE News*, our Congress and Conference Proceedings, and books. The ways in which we promote collaborations is through networking programs, workshops and partnerships with other professional groups, e.g., Engineering World Health.

The IFMBE joins the International Organization for Medical Physicists (IOMP) in a Union called the International Union for Physical and Engineering Sciences in Medicine (IUPESM). These organizations coordinate the ***World Congress on Medical Physics and Biomedical Engineering*** every three years. The next World Congress will be held in Beijing, China in May, 2012.

Happy 50th IFMBE

Herbert F. Voigt, Ph.D., President-Elect IFMBE; Chair, Federation Journal Committee, Chair, Awards Committee

2009 represents the 50th anniversary of the International Federation for Medical and Biological Engineering's (IFMBE's) origin. Congratulations!

It was a pleasure to work with Drs. Paul Fagette and Ratko Magjarević, our Secretary General, on this 50th Anniversary Book, focusing on the last 10 years. Early on we made the decision that the Societies' contributions would be written both in English and in the contributing Society's native language. We feel this portrays and celebrates the IFMBE's International character.

The original objectives of the IFMBE – to encourage research and the application of knowledge, and to disseminate information and promote collaboration within the field of medical, clinical and biological engineering – have not really changed since IFMBE's inception. We continue to support actively regional conferences and for the 1st time in 2006, there was a workshop on progress women are making in Medical and Biological Engineering.

The IFMBE typically works through its Administrative Council, Divisions, Committees and Working Groups. As I assume the Presidency, there are four groups that I am particularly interested in: 1) a new Working Group focused on Genetically Modified Foods, 2) a reactivated Bioethics group, 3) A Global Health Working Group who can address the need to connect the various silos of information and infrastructure available, but are currently isolated from one another, and 4) continued success of the Committee on Women in Medical and Biological Engineering. We are a volunteer organization and I encourage you to consider where you might contribute to our organization's mission. If you have any ideas for creating a new Working Group or reviving one that has become dormant, please contact me.

I am very impressed with the Asian-Pacific activities Working Group. They have created a very nice program that allows young medical and biological engineers to visit each others' labs as a group and then attend the World Congress. I would like to see that effort translated to other parts of the world.

It is very good to see the continued activities of the Clinical Engineering Division. This is of significant interest to the IFMBE and we need to see continued advances in harmonizing the education and training of clinical engineers around the world.

Of course, I am very happy to see the rise in the impact factor of our flagship journal, *Medical and Biological Engineering and Computing* (MBEC) and look forward to the continued leadership of its Editor-in-Chief, Jos Spann.

Finally, I believe it is extremely important to continue the working relationship IFMBE has established with the World Health Organization of the United Nations. The success of this partnership is directly related to the herculean efforts of our Past-President, Professor Nagel.

Herbert F. Voigt
Boston, Massachusetts
USA

A Journey to New Frontiers

Joachim H. Nagel, IFMBE President 2003-2006

When asked back in 1997 whether I would accept a nomination for the Administrative Council elections, I was torn between being proud to be found worthy to serve the Federation in such important office and the apprehension to deprive my Institute of time and efforts without being able to really move a lot in the world of medical and biological engineering. When I finally agreed to accept the nomination, I did not intend to get involved in the administration of the Federation for more than one term. But the first three years of service on the Council completely changed my view. It appeared to me that the Federation did not realize its full potential representing the MBE community and did not get sufficiently involved in the global arena of health and health care though its mission includes the promotion of health and quality of life and the new millennium opened a window of ample opportunities for the flourishing discipline and professions of medical and biological engineering and for the IFMBE to become a global leader not only with regard to science and technology, but also for education, career development, professional issues, public health policy, the provision of health care, patient safety and the application and management of health technologies, including substantial support for developing countries. I became firmly convinced that with sufficient persistence, even a small group of people could make a difference and gain influence on those developments in the world which are related to our discipline and professions. Thus, I embarked on a long journey exploring new frontiers and served the Federation for nine more years as Vice-President, President and Past-President. This is quite a long time for a voluntary engagement, and only due to the unlimited patience and tireless help of my wife Monika, as well as the support from an exceptionally helpful and efficient Secretary General, Professor Ratko Magjarevic, I persevered and our small team could realize the multitude of initiatives thought to reshape and reposition the Federation and our professions.

Being the President of the Federation is an extraordinary privilege, a huge responsibility and an exciting challenge, especially at a time when Medical and Biological Engineering is developing at a breathtaking pace. There is no aspect of our discipline and our professions that is not moving ahead rapidly. Research and education, training and certification, mobility in education and employment, industrial development, ethical issues, health care and health care systems, patient safety, new technologies, international cooperation and support for developing countries are just a few issues that need permanent attention and IFMBE leadership.

To be prepared for the challenges ahead, we needed to optimize the organizational structure of IFMBE as well as communication, making better use of modern technologies to interact and to bring IFMBE officers, council, divisions, working groups and membership closer together. We also had to improve and expand substantially the services to the constituent societies and to all of their individual members world-wide.

Independent of region, country, or the specific field of work, no one should need to ask the question what they are gaining by the membership of their society in IFMBE. Benefits need to be optimized continuously such that they become self-evident to each and every member of the constituent societies. To this end, IFMBE continuously needs to monitor the services we offer, our efficiency and our structure.

Given the enormous amount of administrative work that needs to be done to keep such a large organization functioning, we set up a permanent office to maintain continuity and make the interaction between IFMBE and its members easier. We also developed a new, dynamic homepage, designed for improved communication, offering the possibilities of group work, discussion forums and easier electronic voting through the internet. Another huge and very important step towards improving membership services and benefits was the restructuring of the IFMBE publications, substantially improving the quality and distribution of our scientific journal Medical and Biological Engineering & Computing (MBEC), renewing the book series, further developing the IFMBE News and increasing the number of high quality journals that we are offering to our members free of charge to four: MBEC, Biomechanics and Modeling in Mechanobiology, Cardiovascular Engineering, and Biomedical Microdevices, all released by Springer, the new IFMBE publisher. Through the inclusion of MBEC into various Springer journal packages and the participation in the HINARI Program, set up by WHO together with major publishers, enabling developing countries to gain access to one of the world's largest collections of biomedical and health literature, the readership of our journal could be dramatically expanded from just about 600 library subscriptions to the availability on more than 15 million desktops world-wide. This, of course, resulted in a major boost of publicity and visibility for the Federation.

Finally, we have, in cooperation with the World Health Organization (WHO), embarked on a series of high ranking initiatives aiming not only at globally improved health, health care systems and quality of life, but also at the promotion of the biomedical/clinical engineering professions. It was imperative for us to make every effort to further develop IFMBE as the unified voice of Medical and Biological Engineering, an umbrella for all areas of medical and biological engineering and sciences, including all current directions as well as newly evolving fields, and maintaining the prospects for a bright future of our professions.

Such a large range of activities can, of course, only be funded by a financially healthy organization and I knew that I needed to increase the income of the Federation substantially without raising the membership fees. Re-establishing the IFMBE accounts and re-assessing the value of our journal laid the ground for an extremely positive development of IFMBE finances during my presidency. The new prosperity was mainly due to the recovery of lost income from the past and the fabulous increase of income from the journal that started in 2006, but also, equally important, due to the excellent skills of our treasurer, Professor Shankar M. Krishnan.

Biomedical Engineering has established itself as a key section within the life sciences, a driving force for engineering as well as medical achievements. Creating the professional and public awareness that enables the well deserved recognition for our discipline is a prerequisite for the sustained flourishing of Biomedical Engineering and one of the foremost responsibilities of the IFMBE. During the last decade we have done this very successfully at conferences and meetings that were visited by political decision makers, representatives of the health care systems as well as the medical technology industry and through a large range of new activities and initiatives, above all the active presence of IFMBE in the international work of the World Health Organization (WHO), its general assemblies - the World Health Assemblies - the active participation in WHO Executive Board Meetings, the acceptance as a member of the World Standards Cooperation and the widely visible involvement in various global coalitions for the promotion of health and health care, such as the World Alliance for Patient Safety, the Global Health Workforce Alliance, the Health Actions in Crises Initiative, the participation in the virtual Global Health Professions Network and the prominent participation in the World Health Day 2009 under the theme "Save Lives - Make Hospitals Safe in Emergencies".

IFMBE has entered into a new, close relationship and extensive cooperation with the World Health Organization. In addition to "consciousness raising" through speeches at both the Assemblies and the Executive Board Meetings, our cooperation in numerous projects has brought direct, important results for the Federation, our discipline and our professions, and has helped to make health care safer for the patients. Primum non nocere – first do no harm! Under this motto, attributed to Hippocrates, the World Health Organization with IFMBE as one of its main partners launched the World Alliance for Patient Safety in 2004. WHO, countries, academic institutions, scientific and professional organizations and patients' groups came together from around the globe to advance patient safety by taking a series of key actions to reduce illness, injury and death suffered by patients during medical treatment. Key actions with IFMBE involvement are the so-called challenges "Safe Surgery Saves Lives" and "Technology and Education for Patient Safety".

An important aspect with regard to patient safety is the issue of norms and standards, which were for a long time the almost exclusive concern for the involved industry. In order to strengthen and advance the global voluntary consensus-based standardization system, the World Standards Cooperation (WSC) was established by the International Organization for Standardization (ISO), the International Electrotechnical Commission (IEC) and the International Telecommunication Union (standardization sector, ITU-T). In 2004, the WSC hosted a high-level workshop on International Standards for Medical Technologies in cooperation with WHO, the Global Harmonization Task Force (GHTF), the Association for the Advancement of Medical Instrumentation (AAMI), the European Medical Technology Industry Association (EUCOMED), and the Japan Federation of Medical Devices Associations (JFMDA). As a result of our active participation and subsequent initiatives, IFMBE became a member of the WSC and its Health Care Technology Task Force (HTTF). It has to be considered as a major success for IFMBE to be accepted as a partner by the industry giants as an independent voice of engineering and science within WSC.

Important IFMBE goals are to strengthen the standing of biomedical engineering within the health care systems, enabling biomedical/clinical engineers to take more control over technical, managerial and political decisions, especially furthering their influence and authority with regard to the health technologies, to integrate them stronger into the medical teams in hospitals, and to include them into the group of regulated professions which are responsible for the delivery of healthcare. In order to achieve our goals we established cooperation not only with WHO, but also the International Labour Organization (ILO) and the World Health Professions Alliance

(WHPA). As a major result of these efforts, ILO agreed to classify biomedical engineers not just as engineers as it did in the past, but also as an integral part of the health professionals in its new International Standard Classification of Occupations 2008 (ISCO-08).

Assisting developing countries with regard to health technologies and the necessary clinical engineering workforce has long been an essential activity of the Federation, mainly through its two divisions, the Clinical Engineering Division and the Health Technology Assessment Division. In 2006, we intensified these efforts by teaming up with the International Organization for Medical Physics and establishing a joint Health Technology Task Group (HTTG) under the roof of our umbrella organization, the International Union for Physical and Engineering Sciences in Medicine (IUPESM). The mission of the HTTG is to promote health and quality of life through the advancement of application and management of health technology. In pursuit of its mission the HTTG also promotes international cooperation and communication among those engaged in health-care technology. The Group is in charge of helping countries in defining their needs for health technology and human resources, formulating policies and implementation strategies on acquisition and utilization of appropriate health technology, developing suitable infrastructure for its adequate management and utilization, and identifying and rectifying health system constraints particularly through training and capacity building. For me, this initiative became, next to the patient safety work, one of the most gratifying efforts on behalf of the Federation, and experiencing the enthusiasm as well as the altruism of our volunteers offering their help to less fortunate countries and regions that in many cases have been devastated by catastrophes, is one of the lasting impressions and memories about the time I served the Federation.

Primarily due to the fact that most IFMBE members are working in academic institutions, the Federation has a particularly strong interest in education, training and the promotion of young colleagues. While engaged in offering continuing education and training to clinical engineers in developing countries through its clinical engineering division for decades already, IFMBE never really got involved in the development of educational norms and standards for medical and biological engineering, except for the attempt to establish an international certification and registration program back in 1981, which, unfortunately has not been sustained. Only through the BIOMEDEA Project, which I had the privilege to found in 2004, the Federation ventured into this important area, supporting the BIOMEDEA group of more than 80 mainly European academic institutions which set out to establish Europe-wide consensus on guidelines for the harmonization of high quality Medical and Biological Engineering and Science programs, their accreditation and for the training, certification, registration and continuing education of biomedical engineering professionals working in the health care systems as a first step towards the regulation of clinical engineering as a health profession. This project has gained global attention, and I hope that the Federation will support the expansion of this activity to include all of its member societies.

IFMBE is recognized world-wide as the international representative of Medical and Biological Engineering. This achievement comes with a high responsibility. IFMBE must strive to lead and shape the future of our discipline. The Federation needs to set and defend high standards of our professions, of education, training and certification, publications and of its meetings. As an international federation, it should, in cooperation with WHO and other relevant organizations, lead an effort to develop methods and ways making sure that people in less fortunate and wealthy countries can also benefit from medical progress and good healthcare. As the leading organization in the field, it must guide the profession and explore ways to use new knowledge for diagnostics as well as for the prevention and therapy of diseases, while guarding the ethics of such research and the resulting medical interventions.

During the many years of service for the Federation I noticed the incredibly huge pool of expertise and experience that we have among our members. We need to draw on these resources and further raise public awareness and our influence on professional, societal and political developments. To this end, IFMBE needs to strengthen its activities of generating and publishing Position Statements on the current hot topics in research, education, public health, the healthcare systems, and ethical issues, in short, all matters that are of interest for the MBE community, the public, governments, and funding agencies.

The Federation is well alive and in excellent shape due to an extraordinarily passionate and successful global community of biomedical engineers. I feel very honored that I was given the opportunity to serve as the IFMBE President and I really enjoyed working for this proud Federation. I would like to express my gratitude to the many people who got involved and supported me, both within the Federation and outside.

Commemorative Address for the 50th Anniversary of IFMBE

I have had the privilege of serving as President of the IFMBE for the last 3 years (2006-2009). Assuming the Presidency when it celebrates the 50th anniversary is an extraordinary honor and a huge responsibility. A half century has passed since the IFMBE was founded, and its ability and prosperity to advance the objectives of the profession has not been possible without a very high level of dedication and support of a large number of my predecessors, and for which I am immensely grateful on behalf of IFMBE.

As most of us know, in 1959 a group of medical engineers, physicists and physicians met at the 2nd International Conference of Medical and Biological Engineering, in the UNESCO building, Paris, France to create an organization titled International Federation for Medical Electronics and Biological Engineering. At that time, there were few national biomedical engineering societies and workers in the discipline joined as Associates of the Federation. Later as national societies were formed, these societies became affiliates of the Federation. In the mid-sixties, the name was shortened to International Federation for Medical and Biological Engineering. As the Federation grew, its constituency and objectives changed. During the first ten years of its existence, clinical engineering became a viable subdiscipline with an increasing number of members employed in the healthcare area. The IFMBE mandate was expanded to represent those engaged in Research and Development and in Clinical Engineering. The latter category now represents close to half of the total membership.

The most important milestone in our recent history was the incredible expansion of the needs for medical and biological engineering not only in medical facilities, but also in daily life. Accordingly, the membership in IFMBE is steadily increasing, and currently the Federation has 58 member societies, of which 53 are national and 5 are transnational organizations. This remarkable accomplishment has been undergoing a transformation in regards to the affiliated member organizations. Some of the affiliated organizations are founded on different characteristics in comparison with the traditional national societies. 22 societies do not have direct membership in IFMBE but are members through the European Alliance for Medical and Biological Engineering and Science (EAMBES). EAMBES is an alliance that is an all-inclusive umbrella organization representing all fields of MBES, and includes the European IFMBE member societies, and other national and European societies. The American Institute for Medical and Biological Engineering (AIMBE) is now a representative member of the USA whose member societies represent about 50,000 individuals. These organizations, which include the ICMCC (International Council on Medical and Care Compunetics), CAHTMA (Commission for the Advancement of Healthcare Technology Management in Asia), have also become affiliated with us, and thus the total number of constituent societies of the Federation now is almost 80.

In some countries, clinical engineering seems to be separated from the existing biomedical engineering community. Although there are historical reasons for this, I believe this situation is unfortunate, since each group has much to gain from the other. I thus urge the national societies that are affiliated with the IFMBE and their individual members to reach out to the clinical engineering community so that we can move forward together. In the same context, there are numerous member countries represented by only a small fraction of their professionals through a single organization. The IFMBE should strive to include all of the organizations that are related to biomedical engineering. The IFMBE should work with representatives from individual countries to help formulate a model, appropriate for each country that can be used as a mechanism to bring other biomedical engineering-related organizations under a national umbrella for each country. This would also help in accomplishing many of our Federation's other goals. These issues demonstrate to us that the role of the Federation is definitely changing and should be metamorphosed in accordance with the needs of the expanding scope of biomedical engineering, healthcare technology delivery, and worldwide healthcare organizations.

As President I am well aware of the importance and difficulty in steering such an organization in which the goal is to cover the entire spectrum, all the way from very basic research at a cellular and even subcellular level to the practical application of engineering in the hospital setting and at the bedsides. As important as basic research is, the ultimate purpose of the Federation must be the delivery of healthcare to our fellow citizens and the achievement of the goal of healthcare for all. To meet these needs, the Federation established specialized divisions to address the continuing and long term needs, and a charter has been developed which details the procedures and policies for specialized divisions, committees and working groups of the Federation. These include the Division for Health Care Technology Assessment, and Clinical Engineering, 17 committees (Finance, Federation Journal, Publication & Publicity, Secretaries, Constitution & By-laws, nominating, International Liaisons, Conference, Development, Education & Accreditation, Industry, Membership, Merit Award, Strategic Planning, Web, WHO, and Women in BME committee), and 13 Working groups (Developing Countries, History Book, New Initiatives, Africa/ICHTM, Asia-Pacific Activities, CORAL, Protem Group Europe, Cellular Engineering, Neuroengineering, Physiome, Global Citizen Safety and Security, Physics and Engineering in Oriental and Alternative Medicine, and Bioethics).

The financial situation of the Federation has improved dramatically, and on April 21st 2006 we set up a permanent IFMBE office in the Institute of Physics and Engineering in Medicine (IPEM) in York, England. This should be helpful in maintaining continuity, and making the interaction between the Federation and its member societies easier. These changes have to be optimized so that we are able to disseminate the new horizons of the emerging Medical and Biological Engineering field in a sustainable way as well as support the cultivations of public, national and international recognition of our profession and awareness of its contributions.

As we approach our 50th anniversary, the Federation's leadership will be challenged to utilize and support its experienced and diverse membership to its optimum potential to meet the many opportunities presented by the field of biomedical engineering.

I feel honored that I was given the opportunity to serve as the IFMBE President and I really enjoyed working for the Federation with a proud long history and traditions. Expecting a bright future for IFMBE I am looking forward to sustainable and successful activities in the next half century.

Respectfully submitted,

Makoto Kikuchi
IFMBE President

Tokyo, May 6, 2009

Dov Jaron, President 2000-2003

In contrast with other professional organizations, which are membership based, the IFMBE is a federation that forms an umbrella to unite many national and trans-national societies. Except for a small number of individuals who are members of its International Academy and very few members from countries that do not have an organized society, IFMBE does not have individual members. This presents the first major challenge to the Federation namely, identifying individuals who will devote time and effort to the organization. We have been very fortunate in the past. Many of our best known and most respected biomedical engineers have volunteered their time to the betterment of our profession. As time demands on our colleagues constantly increase, so does the challenge to find both established and young engineers who recognize the importance of working for the Federation. Nevertheless, the tradition of dedication and service, established decades ago, is continuing to be maintained by established members and is augmented by individuals in our member societies who are entering the profession and are learning about the importance of the Federation. As President I would not have been able to serve the Federation if not for the many dedicated individuals who devoted much effort and time to advance the goals of the Federation and the profession of biomedical engineering.

The most important accomplishment during my tenure as President was the acceptance of IFMBE (through IUPESM) as the first engineering focused organization to membership in the International Council of Science (ICSU). This important event signifies the recognition of our accomplishments and contributions to the international scientific community. It brings enormous prestige to individuals whose profession is biomedical engineering.

Another important accomplishment was the establishment of a European organization (The European Alliance for Medical and Biological Engineering and Science) to unite the biomedical engineering community in Europe, elevate the recognition of our profession in this region of the world and provide a strong voice when dealing with the central European government as well as with the many national governments. IFMBE has been a strong advocate of the European Alliance and provided major financial support for its establishment. These efforts were enhanced by a grant from the Research Directorate General of the European Commission. The grant, which was prepared by Niilo Saranummi (past President of IFMBE) with help from myself and other IFMBE Council members, proposed to structure the medical and biological engineering and sciences research sector in Europe; develop tools for coordinating European BME related research and development; and prepare for the future challenges concerning technologies for health. The Alliance has since become a new trans-national organization of the IFMBE. During my presidency I encouraged our other national societies from other parts of the globe to consider forming similar alliances in order to unite and strengthen the profession in their region.

We initiated a Conference Proceedings series. We encouraged all our regional conferences as well as other conferences that we co-sponsor to participate in this series. In addition to increased visibility for IFMBE, we also expected that the Proceedings would provide additional income to the conferences as well as to the IFMBE. The committee on publications prepared a document to provide conference organizers with guidelines on how to take part in the Series.

The Federation Newsletter took on a new appearance thanks to Ratko Magjarevic who became editor of this publication. Ratko was able to attract material for the News, which contained important articles and news items from our various national members.

During the last year of my tenure as President, we made a number of changes to the Constitution and Bylaws of the Federation, all aimed at improving our operations and opening the Federation to new members. One important change was the provision that allows for more than one biomedical engineering organization in a single country to join the IFMBE. This was done to embrace more practicing biomedical engineers than in the past and to allow societies who felt alienated to become part of the international community of BME.

The IFMBE added a new award to recognize individuals who made outstanding contributions to the field. This award was given for the first time in 2003. Therefore, in 2003, the IFMBE was able to bestow the Otto Schmitt Award and the Vladimir K. Zworykin Award. In addition, the IFMBE was able to nominate a biomedical engineer for the IUPESM Merit Award. These awards were presented during the World Congress.

Since the number of face-to-face meetings of the Administrative Council is limited, we approved the use of electronic mail to conduct Council meetings and to vote. This was extended to the business of the General Assembly. On a number of occasions, we have experienced a situation in which a Council member became inactive in the Federation. The new bylaw that was approved, allows for replacement of this individual with a new Council member. It was also determined that societies which do not pay their dues in time would lose their voting rights at the General Assembly. The Federation decided to accept individual membership in countries where a professional society does not exist. Once a Society is formed and becomes a member of IFMBE, such individuals have to forfeit their individual membership and be represented by the society.

An important role of the IFMBE President is to engage in outreach efforts, publicizing the goals and the activities of the Federation. I have been fortunate to be able to make numerous presentations related to the IFMBE activities and tio the biomedical engineering profession on numerous occasions. The topics included:

"Integration of Biomedicine and Engineering: BECON and the New NIH Institute" (Keynote address at the Biomedical Engineering Forum, Tokyo, Japan 2001); "The Future of Biomedical Engineering", (Tsinghua University 2001), "Accreditation of Biomedical Engineering Programs in Europe" (Mediterranean Conference on Medical and Biological Engineering, 2001); "Biomedical Engineering in the USA", (Mediterranean Conference on Medical and Biological Engineering, 2001); "The Future of Biomedical Engineering Research and Education", (Kuala Lumpur, Malaysia June 2002); " Challenges for Biomedical Engineering Research", (Guilin, China May 2002); "The Role of Biomedical Engineering in Biomedical Research and Industrial

Development", (International Symposium on New Medical Equipment Technology, Tokyo, Japan, November 2002, also at a Workshop on Research and Development of Medical and Welfare Equipment and Technology, Fukuoka, Japan, 2002); "E-Health and Global Risk Management", (Korea ASEM Conference on E-Health. Seoul, Korea October, 2003); and"Biomedical Engineering: Leading Research and Health Care Delivery in the 21st Century". (World Engineering Convention, Shanghai, November 2004).

During my presidency of IFMBE, I became active in the International Council for Science (ICSU) and have led since 2002 an initiative on Science for Health and Well Being. Participation in ICSU activities brings our profession to a new level of recognition and prestige. Since its initial participation, IFMBE has been providing important input to the development of a new ICSU program on urban health and most recently I have taken a leading role at the highest level of the organization of the Council with my election to the ICSU Executive Board.

Ratko Magjarević – Secretary General 2003-2009

It was 2003 when I was elected IFMBE Secretary General at the General Assembly in Sydney. A long journey to Down Under, for me from Zagreb, Croatia, two flights, one to Kuala Lumpur lasting 14 hours, and then another lasting additional 11 hours to reach Australia. It was a long journey for the majority of participants of the World Congress (WC) in Medical Physics and Biomedical Engineering, 2003. When the WC site was selected, the representatives of the IFMBE affiliated societies did not know how threatened the event will be. In March 2003, the outbreak of an unusual illness in South Eastern Asia, later called sudden acute respiratory syndrome or SARS, urged many governments to recommend or ever forbid their citizens to travel, especially to far destinations. Though the alert was abolished before the beginning of the Congress, it still affected the number of participants and caused many gray hairs to grow on the heads of the WC organizers. Despite the limitations, more than 2.500 participants managed to attend the Congress and the biomedical community was well represented through the scientific presentations and in all bodies of the Federation that met in Sydney Convention Centre.

Before the World Congress 2003, I was already engaged in the Federation as the Editor of the IFMBE News, the magazine of the Federation which was published bimonthly at the IFMBE web page. I was also asked to help preparing the organization of IFMBE events including the World Congress in Sydney. The dot-com bubble from late nineties already hit most of the members of the BME community and the Congress participants wanted to be connected to the news, to their job and family, read regularly their e-mails… But there was no wireless technology for large number of Internet users at that time, at least not at a reasonable prize. We recognized another opportunity for the Federation to serve its membership by offering free access to Internet and to organize a gathering venue for the members: we hired a booth at the exhibition space, placed into the booth some IFMBE promotional material, primarily on the Federation's Journal *Medical and Biological Engineering and Computing* (MBEC), a table and a few chairs for *ad hoc* short meetings and consultations, and as the most important, a computer with Internet connection. That made the booth a busy place and a meeting point. While the participants were waiting for their turn to connect to the Internet, many of them established international contacts, made appointments and concluded agreements for further collaboration. I can still remember Alan Murray, the Editor in Chief of our journal, *Medical and Biological Engineering and Computing*, standing in the booth and carrying on endless discussions with potential authors and educating young colleagues on how to write a proper scientific paper in BME.

During the Congress, the IFMBE held its regular meeting of the General Assembly which includes elections for the IFMBE Officers and ordinary members of the IFMBE Administrative Council. I was nominated for the office of the Secretary General and I got elected. Of course, I was very happy and honored to have been elected and given such a large responsibility, but at that time I did not realize how much work that office meant and what kind of engagement, including a large number of working hours such a voluntary position in an international association of the size of IFMBE implied. There were two aspects of the increased workload: the first and more important was the need to follow numerous new activities of the IFMBE President, Prof. Joachim Nagel, who took over the office of the IFMBE President during the IFMBE General Assembly, held at the World Congress in Sydney. Prof. Nagel had a plan how to improve the services to the IFMBE membership and the BME Community at large, and Joachim also had an enormous amount of energy which he invested into expansion of the IFMBE activities. Most of those activities remained built into the platform of activities of the Federation after his term of office was over. In this short chapter, reflecting my memories on the period I was serving as Federation's Secretary General, I will not enumerate all activities and achievements of the Federation, since they are covered by the addresses of the Presidents who held the office at that time. I will try to bring my memories, in an anecdotal way, and I hope many of you will be able to refresh your memories and experiences from that period.

Going back to the main track of this presentation, let me speak of the activities of my office. The second cause of increased activities was wide acceptance of Internet for communication and information spreading. The Internet had an enormous impact on communications already since the mid-1990s, but from my experience, the use of electronic mail became a standard in official communications in the IFMBE community about the time when activities for preparing of the World Congress in Sydney began. That also gave a tremendous potential for promotion of the events, but also meant a new organization of the business of the IFMBE secretariat. I use to joke with one of former IFMBE Secretaries in this matter: he actually complained that he was not getting all the information he wanted to get from the secretary general. I reminded him that at the time of his term of office, when the communication between the representatives was by regular mail, he had to prepare and send a letter to the representatives of affiliated societies in spring, in summer he was collecting the feedback, in autumn analyzing the replies and in winter he was informing the Societies on the results of the correspondence. From the early days of my serving as secretary general, the number of received electronic messages was only steadily increasing.

I was very lucky that early after I started serving in the SG office, I was able to hire Ms. Nikolina Volf to help me with the work in the SG Office. Ms. Volf was finishing her studies of English and Italian languages at the University of Zagreb when she started working for IFMBE. Very soon after Nikolina started preparing the regular correspondence for the IFMBE, one of the Committee Chairs reported that my English improved! Nikolina stayed with the IFMBE for more than five years and she was a tremendous help. Belonging to the new "digital" generation, she was supporting me during my travel, by exchange of e-mails or by Skype voice or chat, no matter how large the time difference was between Zagreb and the place where I went to. We are still in touch and Nikolina often advises me or organizers of IFMBE sponsored conferences on how to use EasyChair, the free conference upload and review software recommended for IFMBE conferences.

In order to reduce the pressure to the communication channel to/from the SG Office, we have decided to maximize the use the IFMBE web pages for informing the membership on actions and events related to the Federation. The structure of the web page and the page designer we were using at that time was from early days of Internet as media, but for a number of years we were able to use it very efficiently due to its simplicity. Those web pages were created by my predecessor, Dr. Heikki Teriő, and were managed by a Swedish company. During my first term of office as SG, the IFMBE AC decided to contract the services of a permanent office to our affiliate from UK, Institute of Physics and Engineering in Medicine (IPEM). One of the items in the contract was a re-design of the Federation's web pages, which brought the Federation a much more professional and also eye catching design. Unfortunately, the updating of the information was not so easy as earlier and in some cases, there was a substantial delay in publishing of some information.

From those matters that effected the Federation the most in those six years, I would like to point out switching to a new publisher for Federation's journal. The choice of the new publisher was not an easy task and in negotiations with publishers, Prof. Nagel managed to achieve a large number of privileges for IFMBE individual members. The new publisher, Springer, offered publication of the journal on-line and printed, on paper. The on-line option of MBEC, available from the beginning of 2006, made the journal more attractive for authors and increasing of number of pages per issue as well as publishing the issues on monthly basis proved to be a correct decision. The fact that each member of any IFMBE affiliated society could access on-line and read MBEC free of charge made our journal more visible, more read, more cited. It was and it stayed a success story! On the other hand, Springer made available free of charge to IFMBE members another three peer reviewed journals in biomedical engineering from its production.

Publishing of all publications of the Federation has been entrusted to Springer, and the covers of the Journal, the IFMBE Proceedings and books in the IFMBE book series got a new attractive look, a new brand, such as you can enjoy when taking this Anniversary book into your hands. The IFMBE Proceedings Series was initiated in 2001 and the first Proceedings it the series were those of MEDICON 2001 held in Pula, Croatia. With this first edition, the standards were set and accepted by Springer when they became our publisher. All IFMBE Proceedings since 2006 are published electronically, so that conference participants get them on digital media, and the online version has been made available to IFMBE members under the same conditions as MBEC. The IFMBE Proceedings Series was soon entered into bibliographic databases like WoS ISI Proceedings, Scopus and others. The reputation and stability of the publication means a lot for all those young scientists whose first publication is in conference proceedings.

IFMBE News was one of the publications I was editing for a long time. From a few pages printed from time to time, it grew into a nice bi-monthly publication. There were really nice contributions in the Newsletter, like my favorite, the interviews with Mr. Keith Copeland, legendary IFMBE treasurer and with Prof. Robert Nerem, one of the IFMBE Presidents, both of them prepared by Dr. Monika Nagel.

One of the initiatives that I was especially bound to is introduction of Young Investigators Competition (YIC) to IFMBE sponsored conferences including the World Congresses. An organization like IFMBE, which activities are volunteer based, has to attract the new generation of leaders in the field of biomedical engineering to its activities in an early stage of their careers. In the competition, all young investigators who want to take part in the competition send their papers to a special Awards Committee of the conference, consisting of distinguished researchers from the region. After reviewing the papers, the Committee decides on a limited number of YIC finalists, who then present their contributions (oral presentations or posters) in front of Committee members. They make the evaluation of the presentations and decide on best papers. Diplomas are then presented to the all finalists and the awards the winners of the competition at the closing ceremony. Though I was more a backstage person in most of those selections, I can remember so many happy faces that became red after hearing their names during the winner announcement.

While writing these lines, memories just pop up, and I do have to stop writing at a point. After six years in the office, I was tired from regular workload the SG office brought me, but I had a feeling that those two generations of IFMBE Officers and Administrative Council members have seceded in strengthening and modernizing the Federation. Most of these achievements wouldn't be possible without numerous contributions of biomedical engineers and other colleagues from all over the world. I am very happy that my work for the Federation gave me opportunities to meet so many people dedicated to biomedical engineering and so many nice persons taking care for the benefit of health and wellbeing.

Zagreb, 10 June 2011

Respectfully submitted,
Ratko Magjarevic

IFMBE – Secretary General 1997-2003

I was asked to write a short overview of my six years as Secretary General (SG) of IFMBE. To describe six years of such an activity in few pages is very hard – so many things happened, so many people you met in so many new places. I have written a number of reports and a lot of minutes from meetings, which are stored in the IFMBE archives. These are the formal documents so now I decided to write something that is little more personal, some of my experiences form my years as SG.

The Office of the Secretary General is very important for the Federation and to hold that Office is a challenge, but also a great honour. It is a commission of thrust given by the international biomedical engineering community and very few persons can get it. When I was asked in winter 1997, if I could consider offering myself as a candidate to the Office at the General Assembly (GA) in Nice, I said "yes" without due reflection over how to run the Office and what was actually required from me. Personally, I was in the middle of changing my job – to take over the leadership of the largest Clinical Engineering department of Sweden at the Huddinge University Hospital in Stockholm. That also was a big challenge.

The election in Nice 1997 was not settled before all the votes were counted since there was one more, very strong candidate from USA, Professor Janie Fouke. The vote was quite even; there were 10 votes that differed between the two of us and one vote that was wrongly filled in. I had been active in the board of the Swedish Society 8 years and had represented Sweden in Secretaries Committee and GA in World Congresses (WC) since 1991 in Kyoto. I also had the privilege to get to know a number of persons within IFMBE already during the 80-s through Prof. Jan Persson who was a member of the AC. Still, I was surprised that I got such a support among the delegates of GA since my greatest merit perhaps was my work during the 1993 Nordic Conference in Lund in Sweden. I supported the Administrative Council (AC) in organising their meeting there and I also organised a meeting where the representatives from the Baltic countries, who were especially invited to the conference, could tell us about the actual situation of Biomedical and Clinical Engineering in their countries.

The take over of the tasks of the SG in Nice was not as it should have been since I had to go back to Stockholm the day after GA-meeting because of my new job there. Hence, I was not present at the first AC-meeting so I had to ask my predecessor, Prof. Jos Spaan to take the notes on that meeting even though he probably had planned to do something else.

It was first in March 1998 when the transition of the secretariat physically was finished. At that time all the material, documents, books and correspondence were in Stockholm including a number of boxes containing older IFMBE archives from Ottawa. My plan was to run the Secretary General's Office with a staff of two persons; a secretary/assistant and me. There was a vacant secretary's position at my department and in May 1998, we managed to employ a qualified person. However, very soon I realised that the Department needed more of her time and there was no possibility to cover an additional position for a secretary. Therefore, I decided to do more of the ordinary work for IFMBE by myself.

In 1998 IFMBE still had printed stationery and a booklet directory that was distributed to the National Societies. It was, and I'm convinced that it still is, a great problem to update the information in the directory regardless if it is electronic or not. The problem is to get the actual information from all the national secretaries. Only two weeks after the delivery of the 1998 booklet Directory I received the first notice of errors in it and there was no plan to print another one. The Directory 1998 was the last one that was printed; thereafter the Directory has only been available on the IFMBE web site as a downloadable pdf-file. There was no point using money for printing, because the printed version was always out of date. One of the plans we had in the end of 90's was that IFMBE would have a function on the web site for searching, presenting and administrating an address database. The national secretaries would get access to this database so that they could update the information for their Society. This would certainly have facilitated the communication between the Secretariat and the member Societies as well as the communication between the Societies.

The IFMBE web site was moved from Amsterdam to Sweden in March 1999 and it got a new address, www.ifmbe.org, since it had been before under the IUPESM web address. In the AC-meeting in Seoul 1999 I presented an offer from a professional company to develop the web site. This offer included new layout, a discussion forum for the membership, an electronic meeting place for the committees and divisions, a news site for the membership, a message board, document archive and list-server function. Unfortunately this offer was tabled in Seoul and more offers were taken in late September 1999, but none of them were cheaper than the first one and none of them were accepted. Therefore, I decided to design and make the first step, i.e., the layout, of the new web site by myself. The new web site was launched on the Internet in December 1999.

A proposal for a new layout of the web site was presented for the AC in Pula in 2001. This proposal was made by the same company that hosted the web site and that made the first offer in 1999. Unfortunately, there was no discussion in Pula about this proposal and the Officers reviewed it in their meeting in Sydney in 2002. Unfortunately, too long time had passed from the proposal and the company had gone through a re-organisation and the person who was in charge of the offer at the company had finished. This was a beginning for a long discussion about what the company had promised to do and what they were willing to do. Finally in the autumn 2002 they agreed to make the basic layout according to the discussion the Officers had in Sydney in January 2002. I started to fill the new pages with information in December 2002 and the new site, as it was between 2003 and 2007, was launched in the beginning of June 2003. The total cost for the IFMBE web site from March 1999 when it was moved from Amsterdam to Stockholm to July 2003 was £5456. This sum includes the running cost to December 2003.

IFMBE has the status of Non Governmental Organisation in official relations with UN, WHO and UNIDO. This means that the Federation has the possibility to arrange workshops, seminars, and other activities together with UN organs. We are also invited to take part in UN, WHO and UNIDO meetings where we have the possibility to make remarks and initiate co-operation projects. Every third year the SG of IFMBE writes a report, which is sent to WHO in Geneva for evaluation. In July 2001, I was contacted by the WHO Director of External Cooperation and Partnership, Dr. B.P. Kean, who was requesting of a report of our cooperation with WHO. I had not seen the previous report, because it was written just before I took over the Office. I had one month, in the middle of the summer on me to write this report over the cooperation during the past three years, otherwise we would loose the relationship with WHO. I started to gather information about the common activities and with some help from the WHO liaison person Dr. Andrej Issakov, I managed to finish the 11-page report before the deadline August 31, 2001. Dr. Issakov especially thanked me since the report was written in "WHO style" and he did not have to edit it. The report and the IFMBE NGO status was evaluated at the 109th session of the WHO Executive Board and it decided to maintain the IFMBE in official relations with WHO. In making the decision, the Board commented the efforts of the IFMBE in support of the work of WHO. The Executive Director of the WHO External Relations and Governing Bodies, Dr. A. Asamou-Baah thanked especially IFMBE members for their continuing interest in working with WHO. That was a good base to continue the cooperation.

AC and the Officers meet usually in connection with different conferences and the members have in that way a theoretical possibility to participate in the scientific programme of the that conference. In practice, it is more or less impossible for the SG to be active in the scientific programme. This is especially true during the WC where the Federation has so many meetings of Committees and Working Groups of their own and on top of that there comes also the IUPESM meetings. Organising all these different meetings together with the local organisers is a huge task that takes a lot of time more than one year ahead of the WC. There are so many different wishes concerning the time slot, space and the services during the meeting. The Congress organisers often do not understand that the Federation is running these meeting parallel to the scientific programme and that many of the meeting delegates will have problems with the double bookings with the sessions they are going to chair or with their own presentations. It is a challenge to plan all this and it needs a good deal of patients and capability to compromise.

When the date and time for the meetings has been fixed SG has to chase the many persons who are supposed to report their work or present proposals, so that all the paperwork will be available to the delegates in advance. This is a hope, but we all know that since we are volunteers and we have a number of other obligations there are several documents that are delivered just before the meeting and they should be copied.

I remember asking Marijke Lensing in Nice, who was the Executive Secretary of IFMBE 1991-1997, how she handled the documentations for the World Congress. She told me that all the documentations were copied in Amsterdam and she took it by car to Nice. Thinking of this I decided before the Chicago World Congress in 2000 that I would also copy all the documentation in advance and bring it with me to the USA. What I did not realise was that it would be a huge amount of paper that I was going to have with me in the plane, at the AC-meeting in Evanston and then at the World Congress in Chicago. I had an extra suitcase and four boxes of paper that resulted in overweight charges. I had to pay an additional £177! But, when I arrived in Chicago, the extra suitcase and two of the boxes had disappeared! I never got the boxes back, but the suitcase came to my home address two and half months later, in two pieces. It had been in Florida and Hawaii among other places in the US.

I learned my lesson. Next time, in 2003, I had only one CD with me and I came to Sydney a few days earlier just to have time to find a printer's shop where they could print the paper copies for our meetings. The printing cost in Sydney was only 1/3 of the cost in Chicago.

After the vote for the 2006 World Congress site at the 2000 General Assembly, Korea submitted an appeal to the IUPESM AC. A very intensive exchange of e-mails between the members of the AC started. Korea's appeal stated that the IUPESM Congress Coordinating Committee violated its own procedures during the solicitation of bids and during the selection process. The AC examined the facts and voted that the Korean appeal had sufficient merit to warrant a re-vote. It was then decided that we would start a virtual GA-meeting for the first time in IFMBE history. To begin with, information about the Korean appeal and how IUPESM wanted to proceed with the voting procedure was sent to the 104 GA-Delegates representing each member Society. We received 88 answers to the "Call the virtual meeting to order" in which we first voted on a motion to approve the IUPESM AC recommendation to rescind the vote by the GA and to approve a re-vote for the site selection. The second step in this virtual meeting was to carry out the re-vote of the World Congress 2006 site. We had to send all the information and the ballots to our membership at the same time as the e-mails were sent to the IOMP membership. The procedure started in May 2001 and the count of the votes was finished in mid October 2001. It took a long time from the moment when the whole process started in August 2000 until it was finished. One of the reasons was that during the virtual meeting we allowed 2 weeks time for discussions and 2 weeks for the vote. The bidders also had 3 weeks time for completion of their applications and then 4 weeks for discussion and then the vote. Of course there were some little obscurities in the wording of some information so that some additional weeks went by before the re-vote was actually carried out. It was a very interesting experience giving some indication how the meeting can be held in the future. Using Internet and e-mails for discussions and voting, like in this case, brought about one of the reasons for updating the IFMBE Statues and Bye-Laws that were also done during 2000-2003.

IFMBE is a non-political organisation, but sometimes the global political situation influences our plans. In the beginning of April 2003 the Officers had planned to have their meeting in London just before the Federation Journal Committee (FJC) meeting. However, we found out, just a few days before the meeting that the Vice-president, Prof. Joachim Nagel was not able to travel to London because of the German government's restriction on travel to the UK during the Iraq war. So, we decided to move the meeting to Stockholm and I managed to arrange both accommodations and a meeting room close to the Stockholm airport. After concluding our Officers' meeting, the President, Professor Dov Jaron, Professor Ratko Magjarevic and I could take the plane to London to have our FJC meeting at Savoy Place in London. This was the last Officers' meeting I organised and, in a way, the circle was closed, because the first Officers meeting that I organised in March 1998 was also held in Stockholm.

Back in 1997, after the Nice Conference, I was very enthusiastic about running the secretariat. I cannot say that I would have been pessimistic in 2000 about running the secretariat, but I was realistic. This might sound strange and someone might wonder why I accepted to be re-elected. There are, however, many positive sides to being SG that outweigh the negative sides. I am certain that even the SGs before me have had the same types of problems. People who are not working with the practical issues do not always understand why certain things can become a problem and cause delays.

I went through the old archives in the 12 big boxes, organised, re-marked and re-packed them into boxes that were easier to identify and handle. Most of the archives from my years as Secretary General were stored on a single CD. All the old material, documents, books and correspondence that came to Stockholm in March 1998 were shipped over to Stuttgart for scanning first in January 2006. These archives are complete from the very beginning with correspondence of Vladimir K. Zworykin, Jan Kuipper and others. I wish all the best to the upcoming Secretary Generals with the many important tasks that the Secretary General has.

Heikki Terio, Ph.D. Stockholm March 26, 2010
Secretary General
1997-2003

MBEC after 47 Years

Medical & Biological Engineering & Computing (MBEC), is currently producing its 47th volume implying that soon after the IFMBE was founded, its journal was established. The first Editor-in-Chief was Alfred Nightingale who died too young in 1963 at the age of 40, just before the first issue was launched. A prize, named in his honor, called the Nightingale Prize for "the best scientific paper" published in MBEC was soon established by IPEM in association with the IFMBE.

In the last decade of MBEC's existence, much has changed. The founding fathers thought it wise to create some stability around the journal and so decided that it was best that the publisher, Editor-in-Chief and chair of the Federation Journal Committee (FJC) would all come from the same country. The country of choice was the United Kingdom and the Institute of Electronic Engineering became the publisher. About 30 years ago, I became a member of the FJC and traveled to London twice a year for meetings at the stately IEE building at Savoy Place in London with a view at the Thames. During the passing years many things changed in the scientific world especially related to publishing journals. In the beginning there was a solid base of libraries with significant and stable budgets and, rather importantly, there were no worries about impact factors. Science and engineering, however, grew exponentially and publishing became big business.

IFMBE Officers and Federation Journal Committee meeting, London 2005
(L-R: Shrankhan Krishnan, David Holder, Monika and Joachim Nagel, Steve Smye, Jos Spaan, Alan Murray and Swamy Laxamarayan)

The number of journals increased but at a rate substantially faster than library budgets. The internet made communication between editors, reviewers and authors much easier and the competition grew fiercely. The Impact Factor was invented and the Hirsh factor is becoming rather dominant in judging the success of scientists. The stability that was created by our founding fathers and functioned so well for decades seemed to resist change that was so needed in this new millennium.

With the start of 2006 many things changed. Springer became MBEC's publisher, and the IFMBE chose an Editor-in-Chief from the Academic Medical Center in Amsterdam. There is now an active international team of associate editors in place. The FJC created the necessary atmosphere for the journal to flourish. The results are now becoming measurable.

In 2008, MBEC was cited almost 2700 times and our cited half-life is practically 10. These numbers indicate that we are playing a significant role in the international medical and biological engineering community. The impact factor, IF, is on the rise and jumped in 2008 by 40% to 1.38. The prediction for next year is another significant jump.

As MBEC's Editor-in-Chief I congratulate the IFMBE whole heartedly on its 50th Anniversary and I will do my utmost to maintain and improve the role its journal plays in our scientific and engineering community.

Jos Spaan,
MBEC Editor-In-Chief

IFMBE Publications

Medical and Biological Engineering and Computing 1997-2005

During the period 1997 to 2005 Medical & Biological Engineering & Computing (MBEC) was published by Peter Peregrinus at the Institution of Electrical Engineers in the UK. During this time MBEC appeared every two months. In statistical terms it grew from 506 pages in 1997 to 808 pages in 2005.This was accompanied by a rise in impact factor, from 0.53 to 1.03. This was well below what we aimed for, but was more than made up for by the 'half-life' of papers, which averaged over nine years throughout this period. Put in simple terms, this indicated that you should not look for regular sudden advances, but for research that builds on work published over many years. What more should we expect of a medical engineering journal.

Behind the scenes the journal was managed at the top level by the Federation Journal Committee. We met every six months in London, to discuss how well the journal was doing from both a scientific and financial perspective. This committee was made up of a dedicated and highly supportive group of people, changing over the period, but including Deputy Editors Prof PA Oberg and SW Smye, and members AH McDonald, KT Ison, D Bader, DS Holder, IFMBE Presidents (F Kajiya, JP Morucci, D Jaron, JH Nagel), IFMBE secretaries (JAE Spaan, H Terio, R Magjarevic), other IFMBE representatives (N Saranummi, M Lensing-Kooiman), IEE staff (G Weeler, K Lenton, V Parmar, S Govan, L Baldwin). The committee was chaired by Prof W Bonfield and then by Prof B Brown. It was a joy to work with such a group,

At the day-to-day level, the journal was managed in Newcastle upon Tyne. Newcastle lies in the north east of England on the north bank of the River Tyne, less than 100 km south of the Scottish border, and on the railway line between Edinburgh and London. Newcastle is famous for engineering and railways. George and Robert Stevenson, the pioneers of the steam railway, came from just outside Newcastle, and the gauge for the rail separation comes from the local wagonways. Anyway, back to MBEC. The Editorial Assistants were Frances Mawer and then Anna Newson from 1999 to 2005. They received, recorded and dealt with all manuscripts, and the liaison with referees and authors. This was at times a very demanding task, to make sure that everything flowed as quickly as possible, providing that gentle pressure on referees to put other work aside and devote time to refereeing. Most of the time referee reports agreed remarkably well, even when the referees' background and personal perspectives on the papers differed. Members of the Editorial Board gave significant help with refereeing and with bringing together those referee reports that differed. Inevitably we had to reject papers, but I was often at pains to tell any author that I communicated with that this was just their first step, and that they should listen carefully to referees and others, and develop their work so that it could eventually be published. In all of those dealings, the Editorial Assistants were key. Without them the journal would have ground to a halt.

Initially, papers arrived by post, but we saw an increasing use of email and fax. I still remember some of the early discussions on what the impact of electronic publication might be. It is no understatement to say that the technology underlying journal publication progressed very significantly during this period.

Those who know me will recall that I put effort into helping authors publish their papers, and would often appear at IFMBE meetings to provide publication seminars or workshops or just try to give advice. I had noted that when a paper arrived it was often easy to anticipate whether the referees would accept the paper or not. So how could a 15 minute review allow such a conclusion to be reached? The most important observation was that some abstracts contained no data, or were extremely vague. Some clearly stated an aim, but the results did not correspond to that aim. I always read the abstract thoroughly. Also, some papers betrayed the fact that they may not have been read by any of the supporting authors, even when there might have been four or more. Clearly, none of this was the fault of the often young authors. They lacked training. I hope that the MBEC team in some way contributed to this guidance and training.During my final year, the editor for 2006 was appointed, Prof Jos Spaan. This allowed the new editorial office in Amsterdam to start receiving some of these papers in 2005, ready for publication in the following year. Jos ensured that this worked well, and was ready with a new edition in January 2006.

Let me end on a personal note. There is no doubt that it was hard work. Some of it was tedious. But overall, being Editor brought immense joy and satisfaction. Papers came from all over the world, and many had joint international authorship, some of which was between authors from areas of the world with tension, be this political, religious or ideological. It was great to see how such barriers did not exist when people worked together. I met, or corresponded with, many wonderful people. I learnt much new science and engineering from them. However, for me, I gained much more. The human nature and warmth that makes us work together and share was an experience I will never forget. I want to thank everyone who had any association with the journal in any way. You made my task worthwhile.

Alan Murray
Editor in Chief, Medical & Biological Engineering & Computing 1997-2005

Anna Newson (Editorial Assistant 1999-2005)
and Alan Murray (Editor in Chief 1997-2005)

PART II: Events

World Congresses 1958 - 2015

1958: Paris, France
1959: Paris, France
1960: London, UK
1963: Liege, Belgium
1965: Tokyo, Japan
1967: Stockholm, Sweden
1969: Chicago, USA
1971: Melbourne, Australia
1973: Dresden, Germany
1976: Ottawa, Canada
1979: Jerusalem, Israel
1982: Hamburg, Germany

1985: Helsinki, Finland
1988: San Antonio, USA
1991: Kyoto, Japan
1994: Rio de Janeiro, Brazil
1997: Nice, France
2000: Chicago, USA
2003: Sydney, Australia
2006: Seoul, Korea
2009: Munich, Germany
2012: Beijing, China
2015: Toronto, Canada
2018: Prague, Czech Republic

World Congress Chicago, 2000

The 2000 World Congress was the largest assembly of Medical Physicists and Biomedical Engineers ever held. The Congress incorporated seven organizations: The International Union of Physical and Engineering Sciences in Medicine (IUPESM), The International Federation for Medical and Biological Engineering (IFMBE), The International Organization of Medical Physicists (IOMP), The American Association of Physicists in Medicine (AAPM), The Canadian Organization of Medical Physicists (COMP), The American Institute for Medical and Biological Engineering (AIMBE) and The Engineering in Medicine and Biology Society of the IEEE (EMBS). Close to 6000 participants came to Chicago.

The theme for the Chicago 2000 World Congress was "Global Information Networking for the twenty-first Century". A number of special sessions, supported by United States Federal Agencies were held during the congress. One of them was a symposium on "Balancing Cost with Benefits of Health Care Technologies in the New Millennium: A Global Perspective". This symposium, which focused on the role that technology can play in reducing the cost of health care and the implication of transfer of medical technologies to the developing world, was organized by Dr. Jaron and supported by the National Science Foundation (NSF). The symposium was an outgrowth of a initiative funded jointly by NSF and The Whitaker Foundation on Cost Reducing Health Care Technologies, which was launched in 1992. The symposium incorporated the annual reviews of grantees who were funded by this joint initiative.

One of the highlights of the Congress was the opening ceremony, during which Dr. Francis Collins of the National Institute of Health gave a keynote address titled "The Human Genome Project: A Physics and Engineering Challenge". Dr. Collins outlined the promise of the new discoveries in molecular biology and emphasized the need for participation by engineers and physicists in translating these discoveries to practical applications. Through more than 20 scientific tracks, platform and poster sessions and daily keynote addresses, attendees had an opportunity to learn about the most recent research and to interact with engineering and physics colleagues. They also had an opportunity to enjoy the many attractions of the city of Chicago.

The Congress was preceded by a meeting of the IFMBE Administrative Council. Following the council's deliberations, the members enjoyed an evening picnic at the Ravinia park, followed by an outdoor concert of the Chicago Symphony Orchestra, featuring Daniel Barenboim as conductor and Emanuel Ax as a solo pianist.

Members of the International Advisory Committee to the Congress included representatives from every member national society of IOMP and IFMBE: Ratko Magjarevic, Croatia; Damijan Miklavcic, Slovenia; Makoto Kikuchi, Japan; A.F.M. Verbraak, Netherlands; Tapani Lahtinen, Finland; Nicolas Pallikarakis, Greece; Hiie Hinrikus, Estonia; Annelise Rosenfalck, Denmark; Hermann Gilly, Austria; Marc Nyssen, Belgium; Akos Jobbágy, Hungary; Sun I. Kim, Korea; Jiri Holcik, Czech Republic; Boris Gramatikov, Bulgaria; Hans-Olof Carlsén, Sweden; D. Geiger, France; Antonio Giannella-Neto, Brazil; Tadeusz Palko, Poland; Richard L.G. Kirsner, Australia; Sally

Chapman, Canada; Prodromos Kaplanis, Cyprus; Francis H. Y. Chan, Hong Kong; Guan-Liang Chang, Taiwan; James Goh, Singapore; Shmuel Einav and Dan Adam, Israel; Peter Kneppo, Slovak Republic; Jorge Monzon, Argentina; ZiBin Yang, China; Carlos Hernandez, Spain; Joachim Nagel, Germany; Yorgo Istefanopulos, Turkey.

IFMBE General Assembly at WC2000, Chicago, USA

World Congress Sydney, 2003

The sound of the didgeridoo emerging from the darkness signalled the start of WC2003 – a uniquely Australian beginning for this Congress held on the shores of Sydney harbour. The venue was the recently updated Sydney Convention and Exhibition Centre close to the centre of this vibrant, multicultural city.

When we signed the contract in Chicago in 2000, we never dreamed that we would need to cope with problems such as terrorism, war, a global economic downturn and SARS – all of which led to continually shifting deadlines and more than a few sleepless nights. We got through due to a hard-working and dedicated executive committee with excellent support from our professional conference organiser and continuing international help, good will, encouragement and advice, particularly from Dov Jaron and Joachim Nagel.

We were delighted to welcome so many colleagues and friends from around the world, with over 2200 delegates from over 70 countries, and, in Australia's winter, we were blessed with a week of unbelievably good weather. Personally, having attended all but one of the World Congresses since the first in Hamburg in 1982, it was a dream come true.

Our Keynote and Plenary speakers lived up to their excellent reputations, bringing us the state-of-the-art developments in their fields and pointing us towards the future. From my perspective the two outstanding plenary talks were Dick Kitney's lecture "The Double Helix, and the Role of Engineering and Physical Science in the Post Genomic Age" and Lord May's lecture "The Role of Physics and Engineering in Tomorrow's Medical Education And Practice: The Past May Be A Poor Guide To The Future". In the light of feedback from colleagues and previous World Congresses, we introduced new tracks and changed the thrust of others. As an example, Health Technology Management combined the former Clinical Engineering and Technology Evaluation Tracks and proved to be a great success. Such a broad spectrum of science and technology requires extensive delegation by the Organising Committee to the Track Chairs, who in turn must rely on their Topic Chairs. In the main, this structure worked well, with some of the Tracks being, in effect, international conferences in themselves.

We put a lot of work into good communications. Our e-newsletter *DownUnder 2003* went out every 3 months and we had some editions translated into Chinese and Japanese. Other innovations included the personalised program on the website and the search engine included in the CD of Congress proceedings.

Sheep shearing, sheep dog demonstrations, whip cracking and a koala seemed unlikely starters for the Congress dinner, but they had the diners spellbound. Another "sold out" highlight was an evening at the iconic Sydney Opera House for a performance of Eugene Onegin by the Australian Opera.

Finally, we found that one of the most important spin-offs for the World Congress host country was the participation of locals who would not normally attend international meetings, the enthusiasm that they take with them when they return to their jobs and the effect on the profession at local level. Many of the local societies contributed, through loans, to our seed funding, and this also helped to give them a great sense of pride and "ownership" of the World Congress and a real appreciation of the role of the IFMBE. They also shared, with the IFMBE itself, in the healthy financial profit made by the Congress – an achievement of which we were justly proud, given the difficult world events that at times seemed to be undermining our planning.

Richard Kirsner (WC 2003 Tresurer)

Newly elected 2003 IFMBE Officers and
Administrative Council members

Igor Lackovic during WC 2003
Welcome Reception – meeting a Member of a Koala asylum

World Congress Seoul, 2006

The WC2006 was held from August 27 to September 1, 2006 in Seoul. In WC2006-Seoul, we offered a strong and diverse scientific program covering a wide range of issues and challenges in Medical Physics and Biomedical Engineering on the theme of "Imaging the Future Medicine," that we face today. 2192 members participated from 63 countries, 1160 scientific papers were presented in 246 oral sessions and another 1120 papers were presented in 72 poster sessions covering 25 tracks. In addition to the proffered papers, we opened 18 symposia, 4 workshops, 15 short courses, 4 joint conferences and 14 continuing education courses sponsored by AAMP during the Congress. Ten plenary speakers and more than 100 invited speakers delivered their lectures presenting challenges and opportunities in future developments as well as the latest technical advancements. The scientific program as a whole was designed to promote the fields of medical physics and biomedical engineering to maximize their contributions in improving health care throughout the world. The commercial exhibition was also successful, wherein a total of 207 exhibitors participated. I believe that the congress provided excellent opportunities for communication among medical physicists and biomedical engineers worldwide and contributed to advance physical and engineering sciences in medicine for the benefit of improved health care.

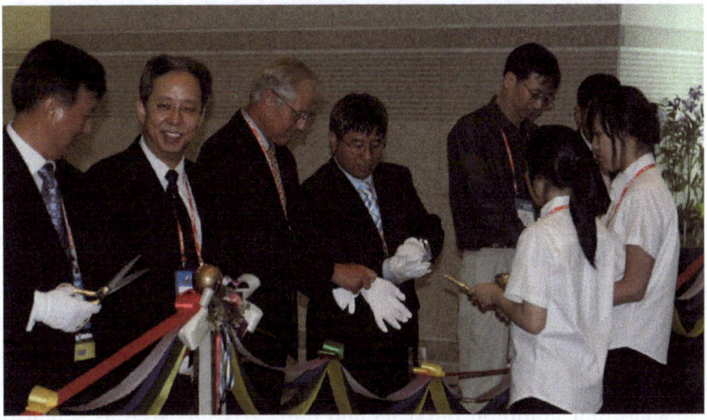

Ribbon cutting at the Welcome Ceremony
in front of the WC 2006 Exhibition hall

When I first attempted to host the World Congress in 1977, at the conference held in Nice, France, no one believed that it would really happen. Many people in Korea said it was too early to host such a big international conference. But I gathered the opinion of the Korean Biomedical Engineering (BME) and Medical Physics (MP) members and got the support from the Korean government. After I lost the bid against Australia, I tried again in the Chicago conference in 2000. The result of the vote was in favour of Switzerland over Korea by 0.06%! However, I found there was a procedural error and I appealed to the IUPESM general assembly. After a significant amount of debate, they decided to revote one year later and finally Korea won the congress.

Many internal and external challenges faced me during the preparation process for the congress. One of the most difficult aspects of organizing the Seoul Congress was coordinating the difference between BME and MP staffs. In Korea, there is not much communication between BME and MP, and therefore we have very different communities and different cultures. Other challenges were the overlap of congress period with the IEEE EMBS annual conference, and the exchange rate of Korean WON against the US Dollar. We lost 25% of our budget due to the steep drop in the exchange rate. But I can proudly say now that we overcame those challenges and the WC2006 was one of the most successful World Congresses ever to be held. I attribute all the success to the members of the organizing committee and our agent, Koconex.

Cultural program – performance by dancers and drummers

Many people say that the primary aim of a scientific conference is to provide an opportunity for researchers to present their latest findings to a wider audience and prompt discussion amongst the community. There is still one unanswered question regarding large conferences in general. Do we still need to physically get together in one place at the same time despite of the wide range of internet access? My answer is YES. Yes, we do need to get together face-to-face. As members of same community, we need to share our common interests and daily lives as well as scientific information in order to promote interaction and communication. In that aspect, the World Congress, as the triennial festival of BME and MP members from all over the world, is all the more important.

Sun I. Kim,
WC 2006 President

World Congres Munich, 2009

Did you enjoy the 11th World Congress on Medical Physics and Biomedical Engineering 2009 in Munich? I really did! Being president of the World Congress together with Wolfgang Schlegel it was quite an extra workload to prepare everything. Not always, but most of time we all together had a lot of fun during preparation and I also remember many valuable experiences that I do not want to miss in my life.

I would like to say "thank you" to Thomas Becks and the team at the VDE who organized everything so perfectly. And a very big „thank you" goes to the theme- and track-chairs. They really had a tough time when the review process started and when the final program had to be assembled. I apologize if we produced some trouble by moving the deadlines. I also apologize if we could not fulfil all the bids e.g. to move some lectures and to add some extra tracks. The program was finally so crowded and to be honest: I am really glad about that!

The congress attracted 3207 scientists from 75 nations of this world. This is the beauty of the World Congress. This is what I really love about it: the peaceful exchange of ideas, the initiation of cross-national and cross-cultural teams who join their knowledge and experiences to push forward biomedical engineering and medical physics. Open discussions in an atmosphere of mutual respect even if you have different opinions about the right way to interpret your results – maybe others could learn from us.

IFMBE Asian Pacific Working Group Meeting during WC 2009

I also enjoyed the World Congress, because it gave me a chance to have a good look into other topics of biomedical engineering and medical physics. Of course I also like to go to specialized conferences where you can go a bit further into depth. But the World Congress offers a chance to get to the top level of other topics very fast and in an enjoyable way. We had 10 plenary lectures, 47 keynote lectures, 207 invited talks, 1621 oral contributions and 838 posters. Especially the plenary talks are what I like – I did not miss a single one. An excellent and enthusiastic scientist reports about the latest results, the open questions and the visions of the future: isn't that great?

Special highlights for me were the two lectures at the opening ceremony. Roger Tsien, the Nobel Laureat who discovered the Green Fluorescent Protein GFP, gave an excellent view into his field of research. I am convinced that there are many exciting challenges that can only be solved by joining the forces of „traditional biomedical engineering" with „cellular engineering" and in particular with labelling of proteins. Completely new ways of diagnosis through integration of optical labelling and optical imaging become visible and on top of that new options of therapy that can target precisely those cells that do not work properly come in reach. I would like to mention that Roger Tsien spent many days on the congress, visited many lectures and often also joined into the discussion. Thank you, Roger Tsien, that was great!

Thank you also to Hermann Requardt. The CEO of Siemens Healthcare gave us an excellent view onto those fields of medical and healthcare systems, where Siemens is already now successful or expects to make profit in future – and I mean this in a very positive sense, because without profit the machine of research and development would not run at all. He pointed out how important it is to always keep the complete chain of healthcare in your mind so that you do not solve local problems that become obsolete already in near future.

I would like to say thank you again to all who have contributed to the success of the congress, and I hope to see you again soon - but latest in 2012 in Bejing!

2009 IFMBE OFFICERS, ADMINISTRATIVE COUNCIL MEMBERS AND CLINICAL ENGINEERING DIVISION BOARD MEMBERS

Regional IFMBE Conferences

Nordic-Baltic Conference

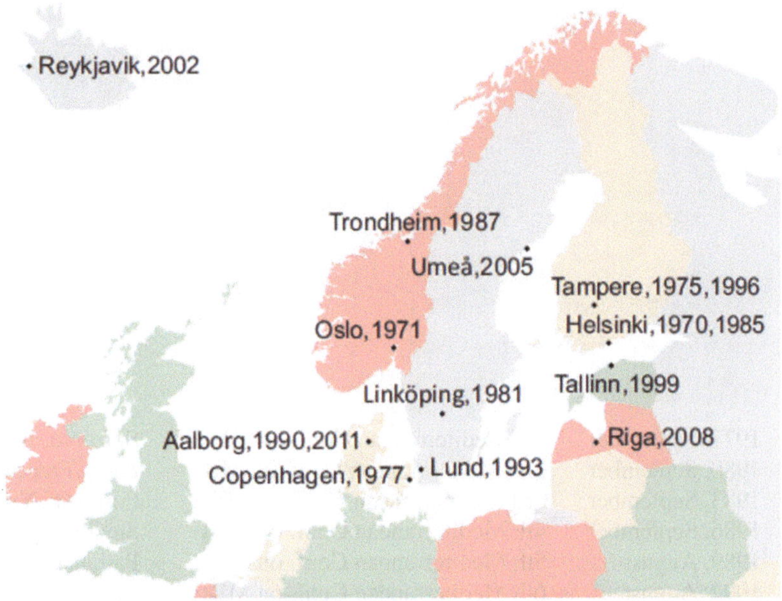

1970, January	1st Nordic Meeting on BME	Helsinki, Finland
1971 June	2nd Nordic Meeting on BME	Oslo, Norway
1975, January	3rd Nordic Meeting on MBE	Tampere, Finland
1977, June	4th Nordic Meeting on MBE	Copenhagen, Denmark
1981, June	5th Nordic Meeting on MBE	Linköping, Sweden
1985, June	6th Nordic Meeting on MBE	Helsinki, Finland
1987, June	7th Nordic Meeting on MBE	Trondheim, Norway
1990, June	8th Nordic Meeting on MBE	Aalborg, Denmark
1993, June	9th Nordic Meeting on MBE	Lund, Sweden
1996, June	10th Nordic Baltic Conf. on BME	Tampere, Finland
1999, June	11th Nordic Baltic Conf. on BME	Tallinn, Estonia
2002, June	12th Nordic Baltic Conf. on BME	Reykjavik, Iceland
2005, June	13th Nordic Baltic Conf. on BME	Umeå, Sweden
2008, June	14th Nordic Baltic Conf. on BME	Riga, Latvia
2011, June	15th Nordic Baltic Conf. on BME	Aalborg, Denmark

Mediterranean Conference

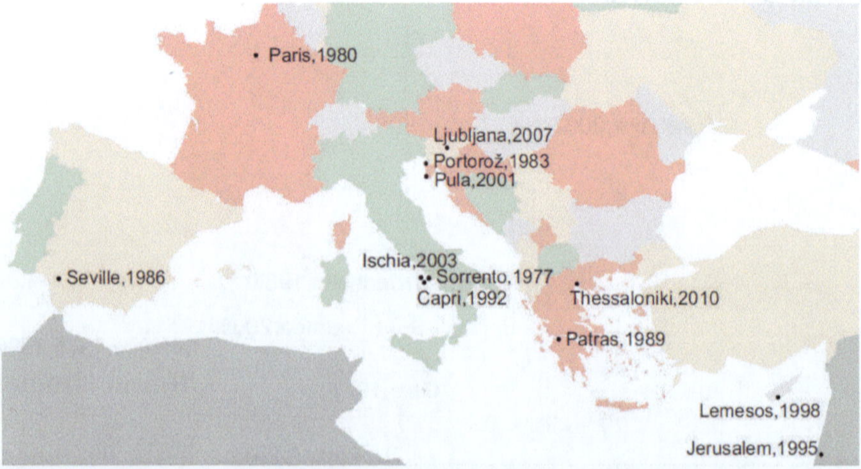

1977, September	1st Mediterranean Conf. on MBE	Sorrento, Italy
1980, September	2nd Mediterranean Conf. on MBE	Paris, France
1983, September	3rd Mediterranean Conf. on MBE	Portoroz, Yugoslavia
1986, September	4th Mediterranean Conf. on MBE	Seville, Spain
1989, August	5th Mediterranean Conf. on MBE	Patras, Greece
1992, August	6th Mediterranean Conf. on MBE	Capri, Italy
1995, September	7th Mediterranean Conf. on MBE	Jerusalem, Israel
1998, June	8th Mediterranean Conf. on MBE	Lemesos, Cyprus
2001, September	9th Mediterranean Conf. on MBE	Pula, Croatia
2004, September	10th Mediterranean Conf. on MBE	Ischia, Italy
2007, September	11th Mediterranean Conf. on MBE	Ljubljana, Slovenia
2010, May	12th Mediterranean Conf. on MBE	Thessaloniki, Greece
2013, September	13th Mediterranean Conf. on MBE	Seville, Spain

Asia-Pacific Conference

1990, October	1st Far Eastern Conf. on MBE	Tokyo, Japan
1993, August	2nd Far Eastern Conf. on MBE	Beijing, China
1996, May	3rd Asia-Pacific Symposium	Taipei, China
1999, September	4th Asia-Pacific Conference on MBE	Seoul, South Korea
2002, December	5th Asia-Pacific Conference on MBE	Mandarin, Singapore
2005, May	6th Asia-Pacific Conference on MBE	Tsukuba, Japan
2008, May	7th Asia-Pacific Conference on MBE	Beijing, China
2011, June	8th Asia-Pacific Conference on MBE	Kuala Lumpur
2014, October	9th Asia-Pacific Conference on MBE	Taipei, China

European Conference

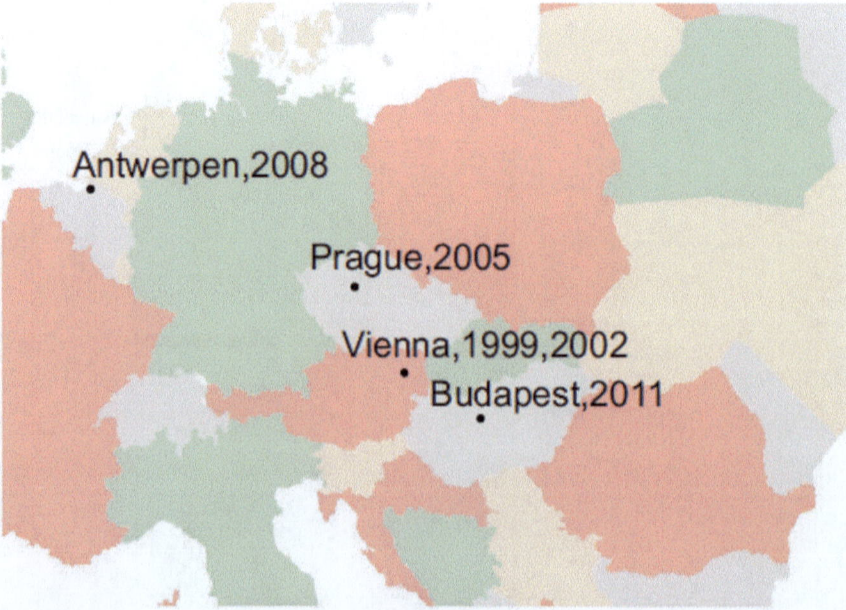

1999, November	1st European MBE Conference	Vienna, Austria
2002, November	2nd European MBE Conference	Vienna, Austria
2005, November	3rd European MBE Conference	Prague, Czech Republic
2008, November	4th European MBE Conference	Antwerpen, Belgium
2011, September	5th European MBE Conference	Budapest, Hungary

Latin American Conference

1998, November	1st Latin American Conf. on BME	Mazatlán, Mexico
2001, May	2nd Latin American Conf. on BME	Havana, Cuba
2004, May	3rd Latin American Conf. on BME	Joao Pessoa, Brasil
2007, May	4th Latin American Conf. on BME	Margarita Island, Venezuela
2011, May	5th Latin American Conf. on BME	Havana, Cuba
2014, October	6th Latin American Conf. on BME	Buenos Aires, Argentina

Cellular Engineering Conference

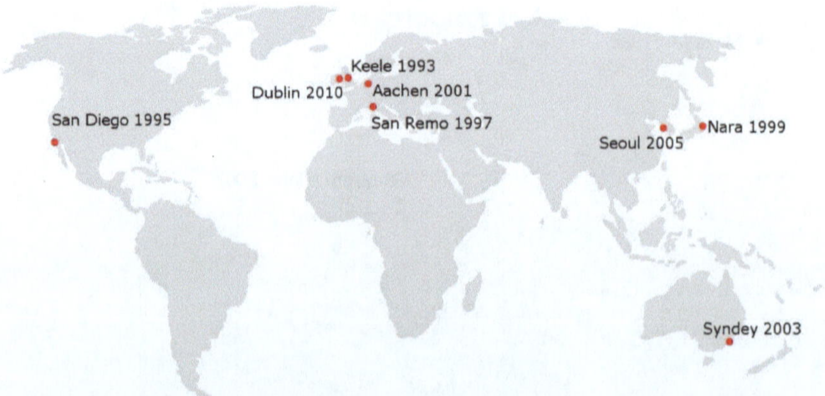

1993, September	1st International Conf. of Cellular Engineering	Keele, U.K.
1995, August	2nd International Conf. of Cellular Engineering	San-Diego, USA
1997, August	3rd International Conf. of Cellular Engineering	San Remo, Italy
1999, November	4th International Conf. of Cellular Engineering	Nara, Japan
2001, July	5th International Conf. of Cellular Engineering	Aachen, Germany
2003, August	6th International Conf. of Cellular Engineering	Sydney, Australia
2005, August	7th International Conf. of Cellular Engineering	Seoul, Korea
2010, June	8th International Conf. on Cell and Stem Cell Eng.	Dublin, Ireland

PART III: Awards

IFMBE AWARDS

Otto Schmitt Awardees:

2009: Graeme Clark
2006: Niilo Saranummi
2003: Basil S. Proimos
2000: Herman P. Schwan

Validimir Zworykin Awardees:

2009: Sune Svanberg
2006: Willem J. Kolff
2003: Per Ake Oberg

Nightingale Award:

2009: Three-dimensional fibril-reinforced finite element model of articular cartilage (LP Li, JT Cheung, W Herzog), Vol 47, Jul, 2009, 607-615

2008: Using a short-term parameter of hart rate variability to distinguish awake from isoflurane anaesthetic states (Hui-Hsun Huang, Yi-Hui, Lee Hsiao-Lung Chan, Yong-Ping Wang, Chi- Hsiang Huang, Shou-Zen Fan), Vol 46, Oct, 2008, 977-984

2007: An effective non parametric method for globally clustering genes from expression profiles (Jingyu Hou, Wei Shi, Gang Li, Wanlei Zhou), Vol 45, Dec, 2007, 1775-1185

2006: Optimal wavelets for biomedical signal compression(Nielsen M, Kamavuako EN, Andersen MM, Lucas MF, Farina D), Vol 44, Mar, 2006, 561-568

2005: Vector projection of biomagnetic fields (L A Bradshaw, A Myers, W O Richards, W Drake, J P Wikswo), Vol 43, Jan, 2005, 85-93

2004: The effects of substrate characteristics on bone cell response to the mechanical environment. (Y Yang, J Magnay, L Cooling, J J Cooper, A Curtis, A J El Haj), Vol 42, (1), 2004, 22-29

2003: Development and validation of implantable sensors for monitoring function of prosthetic heart valves: in vitro studies. (C Lanning, R Shandas), 2003 (41), 416-424

IUPESM AWARDS OF MERIT

2009: Medical Physics: Gary D. Fullerton
Medical Biological Engineering: William Bonfield

2006: Medical Physics: Pedro Andreo
Medical Biological Engineering: Dov Jaron

2003: Medical Physics: Colin Orton
Medical Biological Engineering: Robert Nerem

2000: Medical Physics: Keith Boddy
Medical Biological Engineering: Masao Saito

1997: Medical Physics: Jack Cunningham
Medical Biological Engineering: Robert Plonsey

1994: IUPESM Honours: Rune Walstam (Sweden)

1991: IUPESM Honours: John A. Hopps (Canada)

1988: IUPESM Honours: John Mallard (UK)

Otto Schmitt Award

The First Otto Schmitt Award was given in Chicago. This award is given for "exceptional contributions to the advancement of the field of medical and biological engineering" and the person has to "be a memberof an IFMBE affiliated organisation". The first recipient was Dr Herman Schwan. The award was presented by Dr. Dov Jaron as the chairman of the Awards Committee.

Otto Herbert Schmitt

"... I am confident that a major new quantitative life-based science is emerging which directly, rather than indirectly, seeks its algorithmic mathematical foundations, its discoveries and its analogical applications by emulation of life principles accessible to us via perceptive consciousness insight, rather than by second or third generation laundering through conventional physics, chemistry or other physical sciences that often lead us to conflicting and confusing images of thought."

Otto H. Schmitt (1913-1998) was one of the founders of Biomedical Engineering, a scientist who made a large contribution to the newly emerging fields of Biomedical Engineering and Biophysics and was a great inventor, too. In 1939 he became a faculty member at the University of Minnesota where he had been a teacher in Physics and Zoology at the beginning of his career. By the time he retired in 1983, he was a professor of Bioengineering, Biophysics and Electrical Engineering.

Many of the electronic circuits that are widely used in data acquisition systems and digital systems including computers are a result of his research and development: cathode (or emitter) follower, differential amplifier, chopper stabilized amplifier, the differentiator and integrator circuits etc. Perhaps the most famous among them is a particular digital circuit named the "Schmitt trigger". In a publication in 1938 [1] he described it: "A bistable positive feedback circuit, realizable with vacuum tubes or even fluidic components, but now usually solid-state electronics, which features a selected or adjustable hysteresis band separating initiation of "on" and "off" or "1" and "0" states." He dedicated a lot of his research to exploring the relation between the potential distribution at the body surface and the cardiac current sources, which resulted in a method called "stereo-vector-electro-cardiology". At that time most of the data presentation was planar. The computer rotatable three-dimensional stereoscopic cathode ray presentation he developed was an early attempt at introducing three dimensional medical imaging. He also developed instrumentation for magnetocardiography.

Schmitt formulated the concept of "biomimetics" or the "mimicry of nature" which is based on studying natural processes and Biology and transferring the principles into methods and devices that are usable for mankind. The famous Schmitt trigger was a result of his investigations of the squid nerves. Already on 4 January 1936, the Saint

Louis Globe-Dispatch reported that the "Nerve Study Machine Developed at W. U. by Dr. Otto Schmitt-Holds Possibility of Extending Knowledge of Nervous Systems".

Schmitt's scientific efforts were widely diversified and often multidisciplinary. Though he used to investigate anti-submarine warfare before and during World War II, after the War he dedicated his research to humanitarian matters, primarily to research of new diagnostic methods and the development of medical devices.

He was also very excited with the ideas of advanced health concepts and health care systems supported by technology in order to adjust care individually to the patient and bring it into the patients' home .He was encouraging the introduction of the "Whole Life Personally Portable Medical Record" and using the new technology for quality of life characterization and improvement.

Otto Schmitt was inaugurated into the Minnesota Inventors Hall of Fame and was a member of the National Academy of Engineering. In Chicago in 2000, the General Assembly of the IFMBE established the new Award named after Otto Schmitt. The first awardee was Prof. Herman P. Schwan, who collaborated with Otto Schmitt for a number of years.

Biomimetic Principles

1. Three quadrature compaction algorithm
2. The code-message duality
3. Hierarchical structures of control, regulation, and guidance homeostatic, homeodynamic and heuristically restructurable dynamics.
4. Organismic and social sense of biomimetic principles hierarchically recapitulated.
5. The non-identity of Physics time and biological happening- the H transform and episodal, as against periodic, repetitions.
6. Early efforts toward non-Maxwell electromagnetic field theory in mental and teletransductive phenomena and health features.
7. The universal purpose formulation of negentropic and entropic complementtarity.
8. A set of "principles" of innovation and invention, especially the matrix inversion and mode translation procedures that allow relatively easy "invention to order'.
9. Technological recapitulation of ontogeny.
10. 2 classes of transduction.
11. Interpenetrating domain topology.

Biomimetically Derived Principles

1. The "gradualness" principle by Pavlov. A really new idea takes 15 years to realization, 30 years before it can honestly be reinvented, and a century before it has a secure spot in science.
2. Technology recapitulates Ontogeny much as Ontogeny recapitulates Phylogeny.
3. Biomimetic (or Bioengineering) rules can be used to design and build a new Bioengineering (or Biomimetic) science.

4. By matrix inversion we can direct and accelerate a science toward 'useful goals without hampering its freedom unduly.
5. There are teachable principles of innovation and invention. These need not be occult arts.
6. We need to teach and learn to incorporate the code branch of communication into our science as a co-equal and essential partner to the message branch.
7. A corollary to 6; to be thought of as an entity, a concept must have a name and an identity. We must be prepared to build new mathematical images and procedures in biomimetic form.
8. We must discover culturally hidden scientific taboos that impede progress by traditional unspoken restraints. Consciousness, biological time, fulfilment, spirit, all of these should have their departments in our new Biomimetic Science where they will take on dimensioned and scaled mathematical forms, as for example in the Santosha Index, a co-optimization measure of individual health, happiness and productivity.

Dr. Otto Schmitt at the Blackboard
at University of Stockholm, 1951

Otto Schmitt Awardees

Herman P. Schwan – Otto Schmitt 2000 Awardee

Dr. Schwan is considered as one of the pioneers of biomedical engineering in the United States. His unceasing efforts during the formative years of the profession also had considerable impact globally. He has made enormous contributions to the initiation and development of the field. His most noteworthy services to the profession include: pioneering and continued contributions to the development of biomedical engineering since 1952 through extensive service on national and international societal committees, government boards (NIH consultant and member to several study sections, member GMS Institute Program Project Committee, HEW National Advisory Environmental Health Council) and consultation, and his organisation of unusually successful national and international meetings (1959, 60, 61, 65). He directed one of the first and most successful educational programs in biomedical engineering in the US. Dr Schwan became one of the first biomedical engineers to be recognised for his many research and service contributions and elected to the US National Academy of Engineering. He is a Fellow of the Institute of Electrical and Electronic Engineers (IEEE), the American Association for Advancement of Science (AAAS), and an honorary member of the German Biophysical Society. He received an honorary doctorate from the University of Pennsylvania in 1986.

Herman P. Schwan was born in Aachen, Germany in 1915. He obtained his PhD in Biophysics at the University of Frankfurt in 1940 with distinction, his teaching certificate at the university and his professional doctorate (Dr. Habil) in the fields of physics and biophysics in 1946.

Dr. Schwan worked in 1936-37 and again in 1938 with Telefunken on high frequency and microwave measuring techniques. He became a research associate with the Max Planck Institute of Biophysics in Frankfurt in 1937, an assistant professor with the University of Frankfurt and then associate director of the Max Planck Institute in 1946. In 1947 Dr Schwan came to the United States, working at the Aeromedical Equipment Laboratory of the US Naval Base in Philadelphia. He joined the University of Pennsylvania in 1950, holding in succession various faculty appointments in the Medical School and the School of Engineering. In 1952 he was appointed Head of the Electromedical Division of the Moore School and in 1961 Chairman of the GSAS Group on Biomedical Electronic Engineering.

His best known scientific contribitions include:

- non-ionising radiation; in particular, research on the interaction of elecromagneticfields with biomatter and development of radiation standards of safe exposure to microwaves accepted throughout the western world;
- extensive contributions to present understanding of the electrical and acoustic properties of life matter and development of pertinent precision techniques measurement;

- electrical properties of membranes;
- principles that govern the linear and non-linear properties of electrodes in both frequency and time domain and application of suchprinciples in biomedical research;
- ultrasonic properties of biological systems including tissues, cell suspensions and macromolecules.

Dr Schwan has published almost 300 scientific papers, presented about 350 national and international lectures and received a large number of awards in recognition of his contributions. He has also been recognised for his contribitions to the development of biomedical engineering and for the development of training programs in this field in the United States.

Basil S. Proimos – Otto Schmitt 2003 Awardee

Prof. Proimos received a degree in Mechanical and Electrical Engineering from the Technical University of Athens, Greece, in 1953, the Master of Science degree at the Massachusetts Institute of Technology (MIT), USA, in 1958 and his PhD in Electrical Engineering from the Technical University of Athens in 1965. From 1959 to 1977 he served as Director of the Medical Physics Department at the Greek Anticancer Institute in Athens, Greece, and in 1977 he became Professor of Medicine and Director of the Medical Physics Department at the University of Patras, Greece. Since 1997 he is Emeritus Professor of Medicine at the University of Patras.

Prof. Proimos's research interests were mainly focused on the physics of radiotherapy and the development of conformal radiotherapy techniques using protection of vital organs by absorbers rotating synchronously with the patient (1957-1960) and by gravity oriented devices (1960-2002). From 1994 to 1997 he was coordinator of a "concerted action" of BIOMED I called DYNARAD (DYNAmic RADiotherapy) involving 30 radiotherapy centers within the European Union. Prof. Proimos has published numerous articles on Physics of Radiotherapy in international refereed journals, seven technical and scientific books and he holds two patents for an original radiotherapy unit.

For many years Prof. Proimos was the driving force of European programs promoting education in Medical Physics and Biomedical Engineering not only by teaching engineering and medical students at his university, but as an organiser, coordinator and teacher of a European course on these subjects, taught in English. This program has been offered at the University of Patras for 15 years now and about 30 teachers and 30 new students from European universities participate annually.

In addition, Prof. Proimos was the coordinator of a "Thematic Network" in the SOCRATES Program of the European Union concerning Training and Education for Medical Physics and Engineering Reform in Europe (TEMPERE) involving 38 universities and 10 Scientific/Professional Organisations. He was also the coordinator of a project for Curriculum Development of an Advanced Course (CDA) in Medical Physics and Biomedical Engineering in the framework of SOCRATES involving 30 European universities.

In 1975 Prof. Proimos received the Prize of the "Empiricos Foundation" in Athens and in 1991 the ERASMUS Prize in Brussels, Belgium.

Basil S. Proimos, Ph.D., Emeritus Professor of Medicine, University of Patras, Department of Medical Physics, 265 00 Patras, Greece.

Niilo Saranummi - Otto Schmitt 2006 Awardee

Niilo Saranummi received his Dr.Tech. degree in biomedical engineering at Tampere University of Technology in 1976. He is currently research professor at VTT Technical Research Centre of Finland. His research interests include personal health systems and eHealth, and innovation, technology transfer and technology policy setting in health technologies. He has participated in various roles into ~ 40 EU-funded projects in the area of ICT and health. He is co-founder and chair of HL7 Finland since 1996. He was Editor-in-Chief of IEEE Transactions of Information Technology in Biomedicine (2002-2007). He chairs the International Academy of Medical and Biological Engineering (IAMBE) 2009 – 2012. He chairs IEEE / EMBS's TC on Information Technology for Health. He has served as President of the International Federation for Medical and Biological Engineering (IFMBE), International Union for Physical and Engineering Sciences in Medicine (IUPESM) and European Alliance for Medical and Biological Engineering and Science (EAMBES). He is a Fellow of Finnish Academies of Technology, International Academy of Medical and Biological Engineering (IAMBE), American Institute of Medical and Biological Engineering (AIMBE) and IEEE. He was the recipient of IFMBE's Otto Schmitt award in 2006.

Graeme Clark – Otto Schmitt 2009 Awardee

Clark was born in Camden, New South Wales on August 16, 1935. As a young boy, Clark's father became deaf and this inspired his lifelong mission to help deaf people. Clark attended Camden Primary School and then Sydney Boys High for one year. The remainder of his secondary education was in Royle House at Scots College in Sydney. He then went on to study medicine at the University of Sydney, graduating in 1957 with a Bachelor of Medicine, Bachelor of Surgery (MBBS) with honours and was top of his year. He continued his studies in 1962, in general surgery at the Royal College of Surgeons in Edinburgh.

He then specialised in ear, nose and throat surgery at the Royal National Throat Nose and Ear Hospital and obtained a fellowship in 1969 from the Royal College of Surgeons, London. Clark then returned to Australia and became a Fellow of the Royal Australasian College of Surgeons and in 1969 completed his PhD at the University of Sydney on "Middle Ear & Neural Mechanisms in Hearing and in the Management of Deafness". At the same time, he completed a Master of Surgery thesis on "The

Principles of the Structural Support of the Nose and its Application to Nasal and Septal Surgery'.

In 1976 he returned to England to study at the University of Keele, and to learn more about speech science, as this knowledge was also essential for enabling him to work on converting complicated speech signals into electrical stimuli of the hearing nerve.

Clark considered the notion that hearing, particularly for speech, might be reproduced in people with deafness if the damaged or underdeveloped ear were bypassed and the auditory nerve electrically stimulated, to reproduce the coding of sound. His initial doctoral research at the University of Sydney investigated the effect of the rate of electrical simulation on single cells and groups of cells in the auditory brainstem response, the centre where frequency discrimination is first decoded.

Clark's research demonstrated that an electrode bundle with 'graded stiffness' would pass without injury around the tightening spiral of the cochlea to the speech frequency region. Until this time he had difficulty identifying a way to place the electrode bundle in the cochlea without causing any damage. A breakthrough was achieved during a vacation at the beach – using a seashell to replicate the human cochlea and grass blades (which were flexible at the tip and gradually increasing in stiffness) to represent electrodes.

Clark showed that the electrode bundle had to be free-fitting and the wires terminated with circumferential bands to reduce friction against the outer wall of the cochlea, and so make it easier to pass the required distance. The bands had to be wide enough to minimize the charge density of the electrical current for safety, but narrow enough for localized stimulation of the nerve fibers for the place coding of frequency. In order to satisfy questions about the safety of the device Clark conducted experiments to show that there was a minimal risk of meningitis from a middle ear infection if a fibrous tissue sheath grew around the electrode bundle. The sheath was developed from a connective tissue graft from the person's own body that was placed around the electrode bundle where it entered the cochlea. The ultimate question however was: could speech be coded with multi-channel stimulation so that it could be understood by a deaf person? and this could only be discovered by operating on a deaf person.

The first multi-channel cochlear implant operation was done at the Royal Victorian Eye and Ear Hospital in 1978 by Clark and Dr Brian Pyman. The first person to receive the implant was Rod Saunders who had lost his hearing aged 46.

After successfully completing the surgery Clark in 1978 discovered with his postdoctoral colleague Yit Chow Tong how multi-channel electrical stimulation of the brain could reproduce frequency and intensity as pitch and loudness in severely-to-profoundly deaf adults who originally had hearing before going deaf. Electrical stimulation at low rates of 50 pulses/s was perceived as a pitch of the same frequency, but at rates above 200 pulses/s what was heard was poorly discriminated and a much higher pitch. This discovery established that the timing of electrical stimuli was important for low pitch when this had been difficult to determine with sound. But discrimination of pitch up to 4000 Hertz is required for speech understanding, so Clark emphasized early in the development of the cochlear implant that place coding through multi-channel stimulation would have to be used for the important mid-to-high speech frequencies. Clark and Tong next discovered that place of stimulation, was experienced as timbre, but without a strong pitch sensation. The patient could identify separate sensations according to the site of stimulation in the cochlea.

Clark and Tong at the end of 1978 then made the ground-breaking discovery - the first speech processing strategy to give speech understanding to severely-to-profoundly deaf people using electrical stimulation alone, and in combination with lipreading. The speech processing strategy coded the second formant as place of stimulation along the cochlear array, the amplitude of the second formant as current level, and the voicing frequency as pulse rate across the formant channels.

Clark in December 1978 arranged that his audiologist present open-set words to his first patient, who was able to identify several correctly. Clark realized then that this was the breakthrough in providing speech understanding that everyone had been hoping for "it was the moment I had been waiting for. I went into the adjoining room and cried for joy."

This discovery was established by Clark with objective audiological tests in 1979. The open-set speech test results on this patient were the first time that speech recognition for electrical stimulation alone had been demonstrated, under standardized conditions. Previously single-channel strategies had only shown a small improvement when electrical stimulation was used as a lip reading aid, but no speech understanding for electrical stimulation alone.

As a result Clark went on to operate on a second patient who had been deaf for 17 years. He was able to show that the speech coding strategy was not unique to one person's brain response patterns, and that the memory for speech sounds could persist for many years after the person became deaf.

The successful development of the cochlear implant was confirmed when it gained approval from the Food and Drug Administration (FDA) in the United States. In 1982 Clark supervised the initial clinical studies mandated by the agency, and in 1985 after a world trial the FDA granted approval for the cochlear implant for adults 18 and over who had hearing before going deaf. It thus became the first multi-channel cochlear system to be approved as safe and effective by the FDA or any health regulatory body for giving speech understanding both with lip reading and for electrical stimulation alone in people who had hearing before going deaf.

In 1985, Clark as the surgeon-in-charge performed the cochlear implant surgery on the first children along with Drs Pyman and Webb. The first child was 10 years old and the second was 5 years old. From 1985 to 1990 Clark and the members of his Cochlear Implant Clinic at the Eye and Ear Hospital in Melbourne, followed by other clinics world wide, found that the formant extraction speech coding strategies developed by Clark and team resulted in up to 60% of children being able to understand significant numbers of words and sentences with electrical stimulation alone without help from lipreading. With a strategy that also extracted a band of high frequencies there were increased numbers of children with improved speech perception, speech production and language scores.

In 1990 the FDA announced that the 22-channel cochlear implant was safe and effective in enabling deaf children from ages two through 17 years to understand speech both with and without lipreading. It was the first cochlear implant to be approved by any world regulatory body for deaf children. It is thus the first major advance helping severely-to-profoundly deaf children to communicate in the last 200 years since Sign

Language of the Deaf was developed by l'Abeé de l'Épée at the Paris Deaf School and Speech Reading by Heineke in Germany.

Once Clark established the benefit of a cochlear implant in one ear he commenced research to determine the value of bilateral implants, and an implant in one ear and a hearing aid in the other. His aim was to reproduce the benefits of two ears. These benefits are:

- the ability to localize the direction of a sound (due to differences in the intensity as well as the time of arrival and phase of the sound at each ear),
- hearing speech in noise in each ear due to central neural mechanisms that partially remove the noise, but not the signal (squelch effect or binaural release from masking),
- hearing speech on one side with competing noise in the opposite ear (the head shadow effect),
- loudness summation

(source : http://www.implantecoclear.org)

Elizabeth (left) and Graeme Clark, with IFMBE President, Herbert, and Ronit Voigt, together at the 2009 WC in September in Munich, Germany where Professor Clark received the IFMBE's Otto Schmitt Award

Vladimir K. Zworykin Award

The Vladimir K. Zworykin Award is given to a biomedical engineer for outstanding research contributions in the field of medical and biological engineering. Nominees must be members of an IFMBE affiliated organization. The awardee will receive travel funds to attend the World Congress, where the award will be presented.

The criteria for nominations include innovation and outstanding research contributions to medical and biological engineering.

Vladimir Kozma Zworykin

"I hate what they've done to my child...I would never let my own children watch it."

Vladimir Kozma Zworykin (1889 – 1982), the first President of the IFMBE, is best known to the public through his invention of the cathode-ray tube, at that time called the kinescope, in 1929. Zworykin was one of the first to demonstrate a television system with all the features of modern television tubes. In 1923 he also invented the iconoscope, a tube used in the first television cameras. In the 1930's Zworykin experimented with infrared rays in order to develop devices that could "see" in darkness and in the night. Another application of television technology resulted in the invention of the electronic microscope in 1940 by a team at the RCA Laboratory in Camden, New Jersey, which was led by Vladimir Zworykin. In 1954, he was appointed director of the Medical Electronics Center at Rockefeller University and concentrated his research and development efforts on medical electronic circuits and devices. He was aware of the potential and importance of Biomedical Engineering (at that time more often referred to as Medical Electronics) and therefore worked out a platform for joint operations of scientists and professionals, an international organization in this new science. When the organization was finally founded in Paris in 1959, he became the first and the Founder-President of the International Federation for Medical Electronics and Biological Engineering, today the IFMBE. In his honor, the IFMBE established the Vladimir Zworykin Award in 2000. The first awardee was Prof. P. Åke Öberg who received the award in 2003 in Sydney at the World Congress in Medical Physics and Biomedical Engineering.

Vladimir Zworykin was born in Murom, 200 miles east of Moscow, Russia. He studied Electrical Engineering at the Imperial Institute of Technology in St. Petersburg. His tutor was Boris Rosing, a professor in charge of laboratory projects, who introduced him to experiments of transmitting pictures by wire. They made experiments with a primitive cathode-ray tube that was developed in Germany by Karl Ferdinand Braun. In 1910, they demonstrated a television system using a mechanical scanner in the transmitter and the electronic Braun tube in the receiver. After graduation in 1912

Zworykin received a scholarship and went to study X-rays at the College de France in Paris in the laboratory of a French theoretical physicist Paul Langevin, who was nominated for the Nobel Prize 25 times between 1910 to 1946. After the Russian Revolution, Zworykin emigrated to the United States in 1919, where he first worked at the Westinghouse laboratory in Pittsburgh on the development of radio tubes and photocells. In that period he defended his thesis on photoelectric cells and earned his Ph.D. in Physics at the University of Pittsburgh, Pennsylvania. But his main attention was devoted to the development of television and he patented the iconoscope in 1923 – the first of 120 patents. A little bit later he patented the kinescope, too. In 1929 he was appointed the new director of the Electronic Research Laboratory for the Radio Corporation of America (RCA) in Camden, New Jersey. In the same year, at a convention of radio engineers, Zworykin demonstrated the newly developed television receiver with the kinescope and applied for the first patent in color television.

Vladimir Zworykin is worldwide known as "the father of television" since his fundamental inventions and improvements are the basic technologies of modern, electronic television.

In the early 1930's, Zworykin started working on projects in sensorics, extending the human limits of sensing. It was recorded that he always wanted to "see what you normally cannot see". That research aroused his interest in applications of electronics in medicine. Together with G.A. Morton, he experimented with infrared rays and infrared image tubes in order to develop night-seeing devices. This research enabled the development of first text readers and was used in electric eyes for security systems and garage door openers. Some of the results were also applied to electronically controlled missiles and vehicles. During World War II, he was an advisor of the allies' defense organizations. The other touch with medicine was the application of TV technology to microscopy and improvements in electron multipliers that resulted in development of the electron microscope.

In 1954 he retired from RCA but due to his achievements while working in the company, Zworykin was named their honorary vice president. He was also appointed director of the Medical Electronics Center at Rockefeller University, New York City, where he worked on electronically based medical applications. One of the well known inventions from that period is the radio endosonde, a pill-like transmitter of measurement data from the intestines in 1957. He also developed the ultraviolet color-translating television microscope in 1957. One of the projects he was working on was a telemetric system which transmitted signals from the moving body surface. That system was one of the first applications of transistors in biomedical measurements and though there are no records of a great success of the project, the engineers learned from the difficulties of early transistor applications in instrumentation, particularly medical instrumentation. He was experimenting with electroacupuncture too. One of the conclusions was that the cells under the influence of the electric field are changing their permeability. In the 1970's he acted as researcher at Princeton University and as a visiting professor at the Institute for Molecular and Cellular Evolution, University of Miami, from 1970 until his death in 1982.

In addition to the office of the first President of the IFMBE, he was the National Chair of the Professional Group on Medical Electronics, Institute of Radio Engineers, and the Governor of the International Institute for Medical Electronics and Biological Engineering, Paris.

Most of the numerous awards and honors Zworykin received in his life were related to his most popular inventions i.e. to television. From a very long list, we name only a few:

- Institute of Radio Engineers' Morris Liebmann Memorial prize in 1934;
- American Institute of Electrical Engineers' highest honor in 1952, the Edison Medal;
- University of Liege Medical Electronics Medal, 1963;
- National Academy of Sciences' National Medial of Science in 1967, for his contributions to the instruments of science, engineering, and television and for his stimulation of the application of engineering to medicine.
- He was also named to the National Inventor's Hall of Fame in 1977.

Vladimir Zworykin demonstrates electronic television, 1929
(source: wikipedia.com)

Zworykin Awardees

Per Ake Oberg – Zworykin 2003 Awardee

Prof. Öberg received his M.Sc.E.E. degree in Electronics/Electro-technology at Chalmers Institute of Technology, Göteborg, Sweden, in 1964, and the Dr. of Technology at Uppsala University in 1971. From 1963 to 1972 he was a research associate and Associate Professor at the Department of Physiology and Medical Biophysics of the Uppsala University. In 1972 he joined the Linköping University as a Full Professor of Biomedical Engineering, Head of the Department of Biomedical Engineering (1972-2002) and Director of the Department of Clinical Engineering at the University Hospital (1972-2000).

Prof. Öberg's research interests include biomedical instrumentation, transducers, physiological measurements, biooptics and clinical engineering. He has published more than 450 papers in international peer reviewed journals and at conferences, he holds 30 Swedish and international patents and has co-edited and co-authored 4 books on bioengineering topics. He has served as member of the Editorial Board of several scientific journals, was Editor of the journal of Cellular Engineering and is currently Deputy Editor of the IFMBE journal "Medical and Biological Engineering and Computing". Prof. Öberg is the founding chairman (1979-1983) of the Clinical Engineering Division of the IFMBE, a founding member and chairman (1978-1980 and 1996-1998) of the Swedish Society for Medical Engineering and Medical Physics and a founding member and chairman of the International Academy of Biomedical Engineering. He is a fellow of the Royal Swedish Academy for Engineering Sciences (1980), the Royal Swedish Academy of Sciences (1987), a honorary member of the Finnish Society for Medical Engineering and Medical Physics (1988), a honorary fellow of the Hungarian Academy of Engineering Sciences (1993) and an Honorary Life Member of the IFMBE (2000). He currently is a member-at-large of the IEEE/EMBS Administrative Council.

Prof. Öberg has been the recipient of several prestigious scientific prizes and awards. In 1977 he received the Erna Ebeling Prize of the Swedish Society of Medical Science "for scientific contributions within physiology and biomedical engineering", and in 1981 he was awarded the Fernström Prize of the University of Lund (Sweden) "for outstanding scientific contributions in the field of biomedical engineering". 1982-1983 he taught at the University of Washington in Seattle on a Walker-Ames lectureship award.

Åke P. Öberg, Ph.D, Prof. of Biomedical Engineering, Department of Biomedical Engineering, Linköping University, University Hospital, Linköping S-581085, Sweden.

Willem Johan Kolff – Zworykin 2006 Awardee

Dr. Willem Kolff is considered the "father" of the field of artificial organs, which has led to the modern era of "substitutive medicine". Dr. Kolff's groundbreaking work on the artificial kidney changed kidney failure from a fatal to a treatable disease, prolonging the useful lives of millions of patients. Today, hemodialysis is offered in nearly every country in the world. Willem Johan "Pim" Kolff was born on Valentine's Day (14th February) 1911 in Leiden, The Netherlands. After graduating from Leiden Medical School in 1938, he began postgraduate studies at the University of Groningen as an unpaid assistant under Professor Leo Polak Daniels, director of the University's medical department. Kolff witnesses one of his first patients, a farmer's son, die at the age of 22 of renal failure and hypertension. Kolff, a young father at the time cannot accept that people have to die merely because the body is unable to cleanse the blood sufficiently and starts his research on treating kidney failure, by building a machine that might be able to take over the kidney's function of removing toxins from the blood of patients with uremia. When Germany attacks the Netherlands in May 1940, Kolff founds the first blood bank on the continent of Europe in The Hague. After the suicide of his Jewish mentor Professor Polak Daniels, Kolff decides to leave Groningen rather than to work under the Nazi successor appointed by the Germans. In 1941, he becomes the first internist at the municipal hospital in the small town of Kampen. Kolff grows into a crucial figure in the local resistance. By simulating diseases on people who are in danger of being arrested by the German occupying forces, Kolff succeeds in keeping more than 800 'sick' resistance people and Jews out of the hands of the Nazis. During the darkest days in Nazi-occupied Holland Kolff manages to continue his research on an artificial kidney, working with wooden drums, cellophane tubes, and laundry tubs. He treats 16 patients with his rotating drum artificial kidney in 1943-1944. Fifteen of them die after no more than a few days on his machine and one survives with the kidney having no effect at all on the patient's recovery.

A short time after the liberation of the Netherlands, the 17th patient, 67-year-old Sofia Maria Schafstadt, a Nazi sympathizer, is the first patient whose life is saved thanks to the rotating drum. With her recovery on September 11th, 1945 Kolff launched the field of artificial organs in biomedical engineering research. Kolff receives a PhD in internal medicine *summa cum laude* at the University of Groningen in 1946 based on his thesis about the artificial kidney. In the first post-war years he ships free dialysis machines to researchers in England, Canada and the United States and expands his research to a heart-lung machine (1949). At the age of 39, he emigrates in 1950 with his wife and five young children to the United States, taking a position at the Cleveland Clinic Foundation in Ohio as head of the Department of Artificial Organs and professor of Clinical Investigatio. Here the first pump oxygenator for open heart surgery is used in a patient in 1955, and Kolff devotes himself to the development of the total artificial heart (1957) and the intra-aortic balloon pump (1961). He also improves the techniques of the drum kidney and develops the first disposable twin coil kidney (1955). Together with the Scribner shunt (1960) the invention of Kolff turned kidney failure from a fatal disease into a treatable one.

Under Kolff, the Department of Artificial Organs at the Cleveland Clinic Foundation becomes the most important research center for biomedical engineering in the world at that time. In 1967 Kolff moves from Cleveland to Salt Lake City to direct the Institute of Biomedical Engineering and the Division of Artificial Organs at the University of Utah. Kolff's team of surgeons, technicians and biomedical engineers develops many other types of successful artificial organs, such as artificial eyes, hearing, arms and the subcutaneous peritoneal access device. The permanent total artificial heart (TAH) becomes the most famous one. In 1982, with Kolff at the top of his game at age 71, the Jarvik-7 TAH is implanted in 61-year-old retired dentist Dr. Barney Clark. It is the first successful operation with a permanent TAH in the world. Clark survives for 112 days, after which he dies of pneumoni. Under Kolff's leadership, the University of Utah has since developed one of the world's leading artificial organ research centers.

It is estimated that since the first treatment with the artificial kidney in 1943 in Kampen, Netherlands, more than 20 million lives were saved thanks to Dr. Kolffs inventions and the many people who continue his work. Dr. Kolff has (co-) authored over 600 papers and articles, wrote numerous books, including *Artificial Organs*, and was founding and first president of the *American Society for Artificial Internal Organs* and co-founder of the *International Society for Artificial Organs*. In addition, he has received 123 awards (see insert) and 13 honorary doctorates. He was nominated for the Nobel Prize at least 4 times. Perhaps his most impressive award is the rank of Commander in the Order of Oranje-Nassau, a royal honor bestowed by Queen Juliana of the Netherlands in 1970. His was the first time the award had been presented to an individual living in the United States, and is the highest tribute for accomplishment in the field of science, which can be conferred by the government of the Netherlands upon a native of that country.

The main aim of Dr. Kolff's endeavors as a pioneer of artificial organs has been to restore impaired people to an enjoyable existence. To the question whether it is his intention that a human will ultimately be all composed of artificial organs that prolong his life he replies in a 2001 interview with the magazine *Wisselwerking* (*Interaction*) of the Dutch Organization of Kidney Patients, "This was always my philosophy: If you can give someone a happy life you should do it, if it cannot be a happy life you should not. When someone has an artificial heart and he has enough (of life) he should have a little key, then he can turn it off. But with a code, mind you, so his grandchildren cannot switch him off."

Sune Svanberg – Zworykin 2009 Awardee

Sune Svanberg was born in 1943 in Trollhättan, Sweden. After matriculation exam in Trollhättan in 1962 he started studies of natural sciences at the University of Göteborg, where he received his BSc in 1966. He enrolled the graduate school in physics at Chalmers University of Technology in Göteborg and first spent half a year at the Technical University of Berlin (Prof. H. Bucka) studying atomic resonance spectroscopy. He brought this field back to his university and defended his PhD in this field 1972 (Thesis advisor: Prof. I. Lindgren). After a post-doc year at Columbia University, New York (Prof. W. Happer) and initial work on atomic laser spectroscopy he continued laser-based spectroscopy at Chalmers up till 1980, when he became professor and head of the Atomic Physics Division at Lund Institute of Technology (technical faculty at Lund University). In Lund a vigourous program of laser spectroscopy, including basic atomic physics and applications to energy, environmental and medical research has been pursued. Basic studies include studies of radiative properties of atoms and ions as well as superintense laser/matter interactions (high harmonics generation, X-ray laser pumping and broadband X-ray generation). Applications include laser radar sounding of pollutants in the atmosphere and hydrosphere, laser diagnostics of combustion processes, and laser-based detection and treatment of cancer and cardio-vascular disease.

He has taken the initiative to the formation of three centres for interdisciplinary work now involving more than 40 departments: the Combustion Centre, the Environmental Monitoring Centre and the Medical Laser Centre. He also proposed and helped establish a High-Power Laser Facility, including a multi-terawatt 10 Hz laser. In 1995 he was appointed director of the newly established Lund Laser Centre, which also gained the EC status of a European Large Scale Facility. He has trained a large number of PhD students from home and abroad through the years. He is a member of the Royal Academy of Sciences (and member of its Nobel Committee for Physics) and the Royal Academy of Engineering Sciences. Up till 1995 he was a member of the Swedish National Space Board and the chairman of its Remote Sensing Committee.

He is a Dr honoris causa at the Lund Medical Faculty, at the Science Faculty of University of Latvia, and at the Science Faculty, Université de Liège, a Foreign Member of the Lithuanian Academy of Sciences and the Académie Royale de Belgique, and a Fellow of the American Physical Society and of the Optical Society of America. He has been a member of the Board of Directors of the Optical Society of America and is (together with Claude CohenTannoudji) the recipient of the first EPS Quantum Electronics Prize (1996) and (with Marcus Aldén) recepient of the first Azko Nobel Science Award (1999). He serves on numerous international conference, programme, evaluation and scientific advisory committees. During the years 1987-93 he was a member of the TetraPak Scientific Council and 1993-2000 a member of the Scientific Council of the Volvo Research Foundation. He is the coauthor of almost 500 scientific papers and more than 20 patents and patent applications, and he helped in the formation of several spin-off companies.

Nightingale Award

The first Editor in Chief of MBEC was Alfred Nightingale, a very active man who died too young in 1963at the age of 40, just before the first issue was launched. He-was the then President of the Biological Engineering Society, which subsequently became part of the Institute of Physics and Engineering in Medicine, IPEM, when it was founded in 1995. In 1943, he worked for the British admiralty on electronic underwater sound detectors which may be seen as a precursor of the present day ultrasound technology and joined the physics department of Guy's Hospital in 1947. In honor of this biomedical engineer 'avant la lettre' IPEM in association with the IFMBE has revived the annual Nightingale Prize for the best scientific paper in MBEC. The selection of this paper is in the hands of the Editorial Board of MBEC.

Nightingale 2009 Award

Three-dimensional fibril-reinforced finite element model of articular cartilage

Y Yang, J Magnay, L Cooling, J J Cooper, A Curtis, A J El Haj

ABSTRACT
Collagen fiber orientations in articular cartilage are tissue depth-dependent and joint site-specific. A realistic three-dimensional (3D) fiber orientation has not been implemented in modeling fluid flow-dependent response of articular cartilage; thus the detailed mechanical role of the collagen network may have not been fully understood. In the present study, a previously developed fibril-reinforced model of articular cartilage was extended to account for the 3D fiber orientation. A numerical procedure for the material model was incorporated into the finite element code ABAQUS using the "user material" option. Unconfined compression and indentation testing was evaluated. For indentation testing, we considered a mechanical contact between a solid indenter and a medial femoral condyle, assuming fiber orientations in the surface layer to follow the split-line pattern. The numerical results from the 3D modeling for unconfined compression seemed reasonably to deviate from that of axisymmetric modeling. Significant fiber orientation dependence was observed in the displacement, fluid pressure and velocity for the cases of moderate strain-rates, or during early relaxation. The influence of fiber orientation diminished at static and instantaneous compressions.

Nightingale 2008 Award

Using a short-term parameter of hart rate variability to distinguish awake from isoflurane anaesthetic states

Hui-Hsun Huang, Yi-Hui, Lee Hsiao-Lung Chan, Yong-Ping Wang, Chi- Hsiang Huang, Shou-Zen Fan

ABSTRACT
The measurement of anesthetic depth is important in anesthesiology. Although heart rate variability (HRV) is profoundly affected by general anesthesia, it has not yet been commonly used in this field. One of the reasons is the lack of suitable parameters of HRV for short-term observations. In this study, we designed a time domain parameter of HRV named the similarity index. It was based on observing the trend of the distribution of instantaneous heart rates as time moved. Taking epochs of ECG data as short as 64 s can derive the index. We observed the values of this index of 30 patients when they were awake and under isoflurane anesthesia. The values had very little overlapping between the two states and the prediction probability to distinguish the two states was 0.91. We suggest that HRV, if suitably treated, can play more roles in the monitoring of anesthetic depth.

Nightingale 2007 Award

An effective non parametric method for globally clustering genes from expression profiles

Jingyu Hou, Wei Shi, Gang Li, Wanlei Zhou

ABSTRACT
Clustering is widely used in bioinformatics to find gene correlation patterns. Although many algorithms have been proposed, these are usually confronted with difficulties in meeting the requirements of both automation and high quality. In this paper, we propose a novel algorithm for clustering genes from their expression profiles. The unique features of the proposed algorithm are twofold: it takes into consideration global, rather than local, gene correlation information in clustering processes; and it incorporates clustering quality measurement into the clustering processes to implement non-parametric, automatic and global optimal gene clustering. The evaluation on simulated and real gene data sets demonstrates the effectiveness of the algorithm.

Nightingale 2006 Award

Optimal wavelets for biomedical signal compression

Nielsen M, Kamavuako EN, Andersen MM, Lucas MF, Farina D

ABSTRACT
Signal compression is gaining importance in biomedical engineering due to the potential applications in telemedicine. In this work, we propose a novel scheme of signal compression based on signal-dependent wavelets. To adapt the mother wavelet to the signal for the purpose of compression, it is necessary to define (1) a family of wavelets that depend on a set of parameters and (2) a quality criterion for wavelet selection (i.e., wavelet parameter optimization). We propose the use of an unconstrained parameterization of the wavelet for wavelet optimization. A natural performance

criterion for compression is the minimization of the signal distortion rate given the desired compression rate. For coding the wavelet coefficients, we adopted the embedded zerotree wavelet coding algorithm, although any coding scheme may be used with the proposed wavelet optimization. As a representative example of application, the coding/encoding scheme was applied to surface electromyographic signals recorded from ten subjects. The distortion rate strongly depended on the mother wavelet (for example, for 50% compression rate, optimal wavelet, mean+/-SD, 5.46+/-1.01%; worst wavelet 12.76+/-2.73%). Thus, optimization significantly improved performance with respect to previous approaches based on classic wavelets. The algorithm can be applied to any signal type since the optimal wavelet is selected on a signal-by-signal basis. Examples of application to ECG and EEG signals are also reported.

Nightingale 2005 Award

Vector projection of biomagnetic fields

L A Bradshaw, A Myers, W O Richards, W Drake, J P Wikswo

ABSTRACT
Biomagnetic measurements are increasingly popular as functional imaging techniques for the non-invasive assessment of electrically active tissue. Although most currently available magnetometers utilise only one component of the vector magnetic field, some studies have suggested the possibility of obtaining additional information from recordings of the full magnetic field vector. Three projection techniques were applied to different biomagnetic signals for analysis of the three orthogonal components of the vector magnetic field. Vector magnetic fields obtained from fetal cardiac activity were projected into evenly spaced directions around a unit sphere. The vector magnetic field recorded from multiple intestinal current sources with independent temporal frequencies was then projected. Finally, an external reference signal from an invasive electrode was used to project the recorded vector magnetic fields due to gastric electrical activity. In each case, it was found that the information obtained by examination of the projected magnetic field vectors gave superior clinical insight to that obtained by analysis of any single magnetic field component.

Nightingale 2004 Award

The effects of substrate characteristics on bone cell response to the mechanical environment

Y Yang, J Magnay, L Cooling, J J Cooper, A Curtis, A J El Haj

ABSTRACT
The effect of substrate characteristics on primary human bone cell response to mechanical loading was investigated in this study. The substrates comprised organic and

inorganic materials with a range of hydrophilic and hydrophobic features. Substrate surface topography varied from smooth to particulate to porous. It was found that hydrophilic substrates such as borosilicate glass facilitated bone cell adhesion, in contrast to hydrophobic substrates such as poly(L-lactic acid), in which clumps of cells grew unevenly across the substrate surface. All primary bone cells cultured in the various collagen-coated substrates were responsive to mechanical stimulation. The study showed that, at a low strain level of 1000 microstrain, mechanical stimulation enhanced bone cell differentiation rather than proliferation. Coating the substrates with collagen type I enhanced cell adhesion and promoted an elongated cell morphology, indicating that the presence of specific binding sites on a substrate may be more important than its hydrophilic properties, regardless of the substrate topography.

Nightingale 2003 Award

Development and validation of implantable sensors for monitoring function of prosthetic heart valves: in vitro studies

C Lanning, R Shandas

ABSTRACT
The development of a 'smart' heart valve prosthesis, with the intrinsic ability to monitor thrombus formation, mechanical failure and local haemodynamics and to relay this information externally, would be of significant help to clinicians. The first step towards such a valve is development of the sensors and examination of whether sensor output provides predictive information on function. Custom-made piezo-electric sensors were mounted onto the housing of mechanical valves with various layers of simulated thrombus and bioprosthetic valves with normal and stiffened leaflets. Sensor output was examined using joint time -frequency analysis. Sensors were able to detect leaflet opening and closing with high fidelity for all types of valve. The frequency content of the closing sounds for the mechanical valves contained several peaks between 100 Hz and 10 kHz, whereas closing sounds for the bioprosthetic valve contained energy in a lower frequency range (<1 kHz). A frequency peak of 47 +/- 15 Hz was seen for the normal bioprosthetic valve; this peak increased to 115 +/- 12 Hz for the valve with visibly stiffened leaflets. Total low-frequency (80-3500 Hz) energy content diminished predictably with increasing levels of thrombus for the mechanical valves. Lastly, closing sound intensity correlated well with closing pressure dynamics (dp/dt) ($y = 190x - 443$; $r = 0.90$), indicating that the sensors also provide information on haemodynamics. These studies provide initial evidence regarding the use of embedded sensors to detect prosthetic valve function. Efforts to encapsulate these sensors with telemetry into a custom valve are currently underway.

PART IV: Honours – Academy and Honorary Life Members

International Academy of Medical and Biological Engineering (IAMBE)

The International Academy of Medical and Biological Engineering (IAMBE) is made up of fellows who are recognized for their outstanding contributions to the profession and are elected to the Academy. Nominations are screened by the membership committee of the Academy and the election is by a vote of the Academy fellows.

IAMBE is a part of IFMBE, responsible to the IFMBE General Assembly and its Administrative Council, and operates under the auspices of the IFMBE Constitution and Bylaws. The Academy is established by IFMBE for the following purposes.

The Academy operates under its own Bylaws which are approved by the IFMBE General Assembly. The Academy is to conduct programs which will serve to encourage young people entering the field and their development in the early stages of their career.

Upon request of the IFMBE Administrative Council, the Academy also shall serve as a resource for IFMBE and its various units.

IAMBE Governing Council 2009 - 2012

Governing Council

Niilo SARANUMMI, *Chair*
Roger KAMM, *Chair Elect*
Fumihiko KAJIYA, *Past Chair*
Peter NIEDERER, *Secretary*

Members at Large

Dov JARON
Richard KITNEY
Kit VAUGHAN
Peter HUNTER

Membership Committee

Dov JARON, *Chair*
Jean-Louis COATRIEUX
Shmuel EINAV
Peter HUNTER
Fumihiko KAJIYA
Pai-Chi LI
Jaakko MALMIVUO
Roman MANIEWSKI
Laura ROA
Jos SPAAN
Max VALENTINUZZI

Nominating Committee Chair

Fumihiko KAJIYA
Roger KAMM
Peter NIEDERER (*ex-officio*)

Founding Chair

Robert NEREM

IAMBE Governing Council 2006 - 2009

Governing Council

Fumihiko KAJIYA, *Chair*
Niilo SARANUMMI, *Chair-elect*
Guenter RAU, *Past-Chair*
Richard KITNEY, *Secretary*

Members at Large

Hiee HINRIKUS
Dov JARON
Antonio PEDOTTI
Shoogo UENO

Membership Committee

Peter NIEDERER, *Chair*
Walter CHANG
Shu CHIEN
Jean-Louis COATRIEUX
Peter HUNTER
Roger KAMM
Peter ROLFE
Richard KITNEY
Niilo SARANUMMI
Shoogo UENO
Kit VAUGHAN
Max VALENTINUZZI

Nominating Committee

Guenter RAU, *CHAIR*
Fumihiko KAJIYA
Peter NIEDERER
Robert NEREM
Richard KITNEY (*ex-officio*)

IAMBE Governing Council 2003 - 2006

Governing Council

Gunter RAU, *Chair*
Fumihiko KAJIYA, *Chair-elect*
P. Åke ÖBERG, *Past-Chair*
Niilo SARANUMMI, *Secretary*

Members at large

Richard KITNEY
Robert NEREM
Antonio PEDOTTI
Klaus SCHINDHELM
Yang ZIBIN

Membership Committee

Peter NIEDERER, *Chair*
Shu CHIEN
Jean-Louis COATRIEUX
Ivan DASKALOV
Peter HUNTER
Fumihiko KAJIYA
Robert NEREM
Benno NIGG
Peter ROLFE
Niilo SARANUMMI
Tatsuo TOGAWA
Max VALENTINUZZI

Nominating Committee

Åke ÖBERG, *Chair*
Katsuhiko ATSUMI
Peter NIEDERER
Robert NEREM
Niilo SARANUMMI (*ex-officio*)

IAMBE Governing Council 2000 - 2003

Governing Council

P. Åke ÖBERG, *Chair*
Gunter RAU, *Chair-elect*
Niilo SARANUMMI, *Secretary*

Members at large

Fumihiko KAJIYA
Hiroshi KANAI
Robert NEREM
Max VALENTINUZZI
Peter WELLS

Membership Committee

Robert NEREM, *Chair*
Shu CHIEN
Fumihiko KAJIYA
Peter NIEDERER
Benno NIGG
Robert PLONSEY
Guenter RAU
Peter ROLFE
Niilo SARANUMMI
Klaus SCHINDHELM
Tatsuo TOGAWA
Max VALENTINUZZI

IAMBE Fellows

Founding Fellows

Kazuhiko ATSUMI
Richard S.C. COBBOLD
Floyd DUNN
Leszek FILIPCZYNSKI
Yuan-Cheng B. FUNG
Leslie Alexander GEDDES
Fernand A. ROBERGE
Nándor RICHTER
Guenter RAU
James B. RESWICK
Peter ROLFE
Robert PLONSEY
Annelise ROSENFALCK
Max E. VALENTINUZZI
Peter N.T. WELLS
Masao SAITO
Hiroshi KANAI
Tatsuo TOGAWA
Kazuhiko ATSUMI
Daniel LAURENT
Maciej NALECZ
Robert M. NEREM
P. Åke ÖBERG
John P. PAUL
Klaus SCHINDHELM
Zhenhuang KANG

Fellow Emeritus

Kazuhiko ATSUMI
Uwe FAUST
Floyd DUNN
Leszek FILIPCZYNSKI
Yuan-Cheng B. FUNG
Nándor RICHTER
Basil PROIMOS
James B. RESWICK
Robert PLONSEY
Max E. VALENTINUZZI
Peter N.T. WELLS
Nico WESTERHOF
Hiroshi KANAI
Daniel LAURENT

Jean-Pierre MORUCCI
Maciej NALECZ
John P. PAUL

Fellows inducted in year 2000

Shu CHIEN
Fumihiko KAJIYA
Peter F. NIEDERER
Benno M. NIGG
Antonio PEDOTTI
Léandre G. POURCELOT
Buddy D. RATNER
Niilo SARANUMMI
Colin G. CARO
P-I. BRANEMARK

Fellows inducted in year 2002

Colin ROBERTS
Akira KAMIYA
Jean-Louis COATRIEUX
Zibin YANG

Fellows inducted in year 2003

Lars ARENDT-NIELSEN
Marcello BRACALE
Sergio CERUTTI
Paolo DARIO
Ivan DASKALOV
David DELPY
Hiie HINRIKUS
Peter HUNTER
Helmut HUTTEN
Jaakko MALMIVUO
Roman MANIEWSKI
Richard KITNEY
Toivo KATILA
Robert RENEMAN
Laura ROA
Shunsuke SATO
Christian ROUX
Andrzej WERYNSKI
Erich WINTERMANTEL
Peter KNEPPO

Fellows inducted in year 2005

Nicolas PALLIKARAKIS
Andrew D. McCULLOCH
Einav SHMUEL
Geert W. SCHMID-SCHONBEIN
Christopher L. Kit VAUGHAN

Fellows inducted in year 2006

Yuan-Ting ZHANG
Shoogo UENO
John H. LINEHAN
Roger KAMM
Nozomu HOSHIMIYA
Takeyoshi DOHI
Walter H.-S. CHANG
Ewart R. CARSON
Dov JARON

Fellows inducted in year 2009

Joji ANDO
James B. BASSINGTHWAIGHTE
Ross C. ETHIER
Richard E. KERBER
Yongmin KIM
Pai-Chi LI
Jos A. E. SPAAN
Kazuo TANISHITA

Honorary Life Members

Individuals who have rendered exceptional service to the Federation may have bestowed on them the title of Honorary Life Member.

Honorary Life Members

1971

M. Von ARDENNE
D. H. BEKKERING
P. K. E. DONALDSON
L. E. FLORY
H. W. KNIPPING
M. MARCHAL
A. MONNIER
W. PARIN
W. J. PERKINS
A. REMOND
T. SAKAMOTO
O. H. SCHMITT
B. SHACKEL
C. N. SMYTH
V. K. ZWORYKIN

1976

Jan KUIPER
Jack A. HOPPS
Ryosei KASHIDA

1979

David DEWHURST
Dennis HILL

1982

Lester GOODMAN
Joseph WEINMAN
Kenkichi SUHARA

1984

Zoltan KATONA

1985

Keith COPELAND
Masao SAITO

1988

David SIMPSON

1991

Øivind LORENTSEN

1994

Hiroshi ABE
Leslie GEDDES
Orest Z. ROY

2000

Kazuhiko ATSUMI
Daniel LAURENT
Maciej NALECZ
Nandor RICHTER
Annelise ROSENFALCK
Åke ÖBERG

2003

Fumihiko KAJIYA
Richard KIRSNER
Niilo SARANUMMI

2006

Andrei ISSAKOV

PART V: Governing Bodies – Officers and AC Members

IFMBE Officers

The officers of the Federation shall be:

a) the President,
b) the Vice-President,
c) the Secretary-General, and
d) the Treasurer

IFMBE Administrative Council

The affairs of the Federation shall be administered by an Administrative Council, whose voting members are:

a) the officers,
b) the immediate Past-President, and
c) eight members, elected by the General Assembly from the membership of the Member Organisations.

The Editor, the Deputy Editor, and all Committee Chairmen may attend Administrative Council Meetings if they so wish. The President and the Administrative Council may both invite individuals to attend meetings of the Administrative Council. The Administrative Council shall have full power to act on behalf of the Federation and to take such action in all matters of administration and policy, including conference policy, not expressly reserved by these Articles for the General Assembly, but in all its actions shall be responsible to the General Assembly. A meeting of the Administrative Council shall be convened by the President or by a petition of at least thirty per cent of the voting members of the Administrative Council. The business of the Administrative Council may be carried out by postal of electronic mail, as specified in the Bylaws.

Administrative Council 2009- 2012

President

Herbert VOIGT

Vice-President (President-Elect)

Ratko MAGJAREVIC

Secretary General

James GOH CHO HONG

Treasurer

Shankhar MUTHU KRISHNAN

Immediate Past-President

Makoto KIKUCHI

Ordinary Members of the Administrative Council

Said Jorge CALIL
Ákos JOBBÁGY
Sun I. KIM
Luis KUN
Olof LINDAHL
Jan WOJCICKI
Marc NYSSEN
Guruprasas MADHAVAN

Administrative Council 2006– 2009

President

Makoto KIKUCHI

Vice-President (President-Elect)

Herbert VOIGT

Secretary General

Ratko MAGJAREVIC

Treasurer

Shankhar MUTHU KRISHNAN

Immediate Past-President

Joachim NAGEL

Ordinary Members of the Administrative Council

Said Jorge CALIL
James GOH CHO HONG
Ákos JOBBÁGY
Sun I. KIM
Depei LIU
Teikki TERIO
Katsuhiko TSUJIOKA
Jan WOJCICKI

Administrative Council 2003 – 2006

President

Joachim NAGEL

Vice-President (President-Elect)

Makoto KIKUCHI

Secretary General

Ratko MAGJAREVIC

Treasurer

Shankhar MUTHU KRISHNAN

Immediate Past-President

Dov JARON

Ordinary Members of the Administrative Council

Antonio Fernando Catelli INFANTOSI
Didier GEIGER
Marc NYSSEN
Myoungho LEE
Depei LIU
Heikki TERIO
Swamy LAXMINARAYAN
Ofer BARNEA

Administrative Council 2000 – 2003

President

Dov JARON

Vice-President (President-Elect)

Joachim NAGEL

Secretary General

Heikki TERIO

Treasurer

Mladen POLUTA

Immediate Past-President

Jean-Pierre MORUCCI

Ordinary Members of the Administrative Council

Joe C. BARBENEL
Antonio Fernando Catelli INFANTOSI
Didier GEIGER
Walter CHANG
Janie FOUKE
Marc NYSSEN
Myoungho LEE
Jos A. E. SPAAN

Administrative Council 1997 – 2000

President

Jean-Pierre MORUCCI

Vice-President (President-Elect)

Dov JARON

Secretary General

Heikki TERIO

Treasurer

Mladen POLUTA

Immediate Past-President

Famihiko KAJIYA

Ordinary Members of the Administrative Council

Antonio Fernando Catelli INFANTOSI
Helmut HUTTEN
Joachim NAGEL
Yang Zi BIN
Joe C. BARBENEL
Walter H. CHANG
Janie FOUKE
Jos A. E. SPAAN

PART VI: Member Societies

American Institute for Medical and Biological Engineering (AIMBE)

American Institute for Medical and Biological Engineering

Congratulations on the 50th anniversary of IFMBE!

The American Institute for Medical and Biological Engineering (AIMBE) was founded in 1991 to establish a clear and comprehensive identity for the medical and biological engineering field—which is the bridge between the principles of engineering science and practice, and the problems and issues of biological and medical science and practice.

AIMBE's headquarters in Washington, D.C. organizes forums for professional exchanges, keeps the community abreast of new developments in the field, and advocates to the U.S. government for policies benefitting the medical and biological engineering. Our members include Fellows, the top 2% of the field, Academic Institutions, Industry and other societies engaged in medical and biological engineering.

AIMBE's priorities include increasing funding for research and development, safely bolstering the rate of innovation by removing barriers and building infrastructure, creating a more diverse engineering workforce, and improving general awareness of how medical and biological engineering improves public health.

AIMBE recognizes the value of collaboration, particularly at the international level, and shares IFMBE's commitment to ethically use science, engineering and technology to benefit people across the world

We are proud to be affiliated with IFMBE and look forward to another 50 years of progress in medical and biological engineering and therefore improvement of the human condition.

Sincerely,

John Thomas Watson

John T. Watson, Ph.D.
President

Sociedad Argentina de Bioingeniería (SABI)

MEMORY Years 2007-2009

This report highlights the following as significant events.

• SABI Congress 2007

The XVI Argentine Congress of Bioengineering and the V Conference of Clinical Engineering (SABI2007) were held in San Juan from 26 to 28 September 2007, organized by the Regional Cuyo Argentina Sociedad de Bioingeniería and the School of Engineering at the Universidad Nacional de San Juan.

SABI2007 Congress was attended by 524 participants from Chile, Peru, Uruguay, Brazil, Colombia, United States, Spain, Italy, Germany and Argentina. The congress was sponsored by the Government of the Province San Juan, the National Agency for Promotion of Science and Technology (ANPCyT), the National Council of Scientific and Technical Research (CONICET), the Institute of Electrical and Electronics Engineers (IEEE) Section Argentina and by different local and national firms.

A total of 186 scientific contributions were received, of which 152 papers were finally accepted. There were 7 plenary lectures by leading foreign researchers from Chile, Spain, United States nationals.

Under SABI2007 activities, a contest was held to reward the best work of undergraduate and graduate students. Twenty (20) papers were submitted to the undergraduate competition and 23 to the graduate contest.

A day before the congress, a SABI Student Day was held. This was attended by 214 students, from Entre Rios, Mendoza and San Juan, teachers, alumni and general public. A total of 17 papers were presented in different sessions and a round table was conducted, in which several bioengineers shared their work experiences in various fields outside the university related to work in bioengineering

• 1st Special Issue of the Journal of SABI
The first special issue of the Journal was published in 2007. In this case, it was about Clinical Engineering, in addition to the biannual edition with a circulation of 500 copies per number. The special issue was coordinated by Ing. Eduardo de Forteza, Chapter President SABI Clinical Engineering.

• XII International Conference of Clinical Engineering and Medical Technology.
Organized by the Group of Studies in Clinic Engineering (GEIC) of the National University of Entre Ríos (UNER), this conference was held in Paraná city, Entre Ríos, from 20 to 22 August of 2008. A total of 206 professionals and students from 7 countries in Latin America attended this Conference. Within this framework, there were 37 oral and 20 poster presentations and 3 training Workshops, the latter with full attendance quota. These Workshops were sponsored by associations like the American

College of Clinical Engineering (ACCE), the Peruvian Association of Clinical Engineering (APIC), the Argentine Committee Maintenance (CAM), the Chamber of Hospital Equipment Manufacturer in Argentina (CAEHFA), Latin American universities and medical equipment firms.

• New website SABI

The society's web page was transferred to the administration of the Regional Rosario, the Web master being Mario Figge (mario@3tristestigres.com). The web page is www.sabi.org.ar

• 2nd Special Issue of the Journal of SABI

In addition to the biannual edition of the SABI Journal, a 2nd Special Issue was published in 2008, with an edition of 500 copies per issue; in this case, it dealt with Biomechanics and Rehabilitation Engineering. This special issue was coordinated by Msc Bioing. Taberning Carolina, Msc. Bioing. Ruben Acevedo and Dr. Bioing. Ariel Braidot.

•BioRed Meeting

In accordance with the Agreement Act "Academic, Science and Technology Cooperation" for degree studies of Bioengineering and Biomedical Engineering for the six (6) Academics Units in the country, representatives of these careers met at the city of Cordoba in July 2008. There were coordinated activities of the Exchange Subproject 2008-2010 Multi-Year Plan Improvement Project in Engineering Education (PROMEI II). Strategies were stated for support research systems, professional incumbencies, and for participation in state agencies such as the ANMAT.

The meeting had 2 guests: The Director of Medical Technology from ANMAT and the President of the Professional Council of the province of Cordoba.

•2nd Argentine Conference on Brain Computer Interfaces JAICC 2009.

The Rehabilitation Engineering Laboratory and Research in Neuromuscular and Sensory (LIRINS), from the Faculty of Engineering UNER, together with the "regional litoral de SABI", organized the 2nd Argentine Conference on Interfaces Computer Brain from 23rd and 24th April 2009 in Paraná, Entre Rios.

There were not only works from Argentina, but also from Chile, Colombia and Mexico, which gave a Latin American character to the event. There were prestigious guests, national and international, from Switzerland, France, Mexico and Chile.

• Agreement with SADIO

In August 2009, an agreement was signed between the SABI and the Argentinean Society of Operative Informatics (SADIO). A Liaison Committee was also conformed with other medical societies such as the Argentinean Medical Association (AMA) and the Argentinean Medical Informatics Association (AMIA). The scope of this committee is to coordinate joint activities.

• SABI Congress 2009

In October, 14-16, 2009, the XVII Argentinean Congress of Bioengineering and the VI Conference of Clinical Engineering (SABI2009), were held at the Centro Cultural Bernardino Rivadavia (CCBR), in the City of Rosario.

These meetings were organized by the Rosario branch of the Argentinean Society of Bioengineering and the Polo Tecnológico de Rosario. They were sponsored by the National Research Council of Scientific and Technical Research (CONICET), the Argentinean section of the Institute of Electrical and Electronics Engineers (IEEE), the Deputies Chamber of the Government of the province of Santa Fe, the American Open University, the Spain International Cooperation Agency (AECI), the Drugs, National Agency of Food and Technology (ANMAT) and by several national and international private firms.

A total of 750 people attended the congress, 203 scientific contributions were submitted, and finally 183 papers were accepted. Several speeches were lectured by 10 prominent foreigner researchers from Uruguay, Chile, Brazil, Paraguay, Peru, Venezuela, Colombia, Cuba, Spain and USA.

Two different prizes were awarded to the best paper in the undergraduate and graduate categories, respectively. Simultaneously, the SABI student workshop were developed. A total of 14 papers were submitted and 13 were finally accepted, all coming from the University of Tucuman, San Juan and Entre Rios. These Sessions were conducted with full participation of students from the academic units in the area of Bioengineering and Biomedical Engineering in the country.

• Creation of the Engineering in Rehabilitation Chapter

During the last SABI 2009 Congress and, as decided by the General Members Assembly, the Chapter of Engineering in Rehabilitation was established. It was founded by those Society members who carry on research and development in reahabilitation. The members of this chapter designated Dr. Ing. Juan Pablo Graffigna as a chairman, discussed the basic objectives and defined their organization and operating mode.

• Regular Assembly SABI

On the 15 October 2009, as part of the activities of the SABI Congress 2009, the Ordinary Meeting of the Society Argentina of Bioengineering was held at the Centro Cultural Bernardino Rivadavia from Rosario. Among the agenda items of the day, the following were discussed: Processing and approval of the report, inventory, balance and report by the Review Committee of Auditors in the periods 01/08/07 to 31/07/08 and 01/08/08 to 31/07/09, and new authorities of the Society.

The Overseer Lawyer of "Personería Jurídica" Mrs. Manso attended the Assembly and 63 members were present, who signed the attendance book.

The memory, balance, and inventory were approved and the outgoing Directive Committee raises a proposal for a Directive Committee, which was approved. The committee was conformed as follows:

PRESIDENT:
Ana Maria Echenique amechenique@gateme.unsj.edu.ar)

VICE-PRESIDENT:
Ariel Braidot (abraidot@bioingenieria.edu.ar)

SECRETARY:
Virginia Ballarin (vballari@fi.mdp.edu.ar)

TREASURER:
Myriam Herrera (mherrera@herrera.unt.edu.ar)

1st MEMBER:
Juan Pablo Graffigna (jgraffig@gateme.unsj.edu.ar)
2nd MEMBER:
Dario Rocha (dario@consultar.org)
Substituting 1st MEMBER:
Juan Manuel Olivera (jolivera@herrera.unt.edu.ar)
Substituting 2nd MEMBER:
Diego Beltramone (dbeltramone@efn.uncor.edu)
Accounts Reviewer:
Adriana Scandurra (scandu@fi.mdp.edu.ar)
and Ernesto Federico Treo (etreo@gateme.unsj.edu.ar)

At the end of the Assembly, the Mar del Plata Regional is confirmed as the organizer of the next SABI Congress 2011.

On the Closing Act of the SABI Congress 2009, the New Directive Committee makes their proposals known, which are the following:

• Create strategies that enable its active participation as a national referent, thus facilitating communication with key stakeholders. One of the strategies is to propose the creation of an Advisory Council to the Directive Committee of SABI with representatives of strategic areas such as the Secretariat of Science and Technology, International Organizations, CONICET, CONEAU, etc..
• Propose before the Administrative Committee CORAL the possibility of designating the Faculty of Engineering of the Universidad Nacional de Entre Ríos (UNER) as a potential host institution of the XXI Meeting of CORAL in August, 2010
• Propose and set up committees to work on different lines with clear well-defined objectives, allowing greater participation of all partners; for example, a committee of technological development with the participation of the various branches, for the purpose of encouraging the development of projects of technological basis in bioengineering.
• Organize regional programmes with an initial proposal of a rotating regional activity on a defined topic and transmit this to the rest, for example, through teleconference. Promote activities (workshops, symposia, exchange of researchers between different centres, etc.) of the chapters created (Clinical Engineering and Rehabilitation Engineering) and document them.
• Establish a SABI Publishing committee, with a coordinator for editing not only the magazine but also books with educational objectives and / or scientists from the various issues relevant to Bioengineering.
• Achieve the journal indexing, generating the proper format, for which, a formal commitment is required of all the branches for sending papers which will increase the number of annual reviews. Explore the possibility of parallel editing in digital format, which lets us provide a complete index and the first page of the article.
• Promote joint activities with other scientific societies, with which an agreement has been formalized (e.g., Operational Informatics Argentine Society SADIO).

• Perform monthly virtual meetings of the members of the Directive Committee and of regional delegate members
• Generate an electronic database of partners, with a search engine smart (research topics, regional, etc).
• Evaluate the development chapter of Medical Informatics, recently proposed.

Sociedad Argentina de Bioingeniería

MEMORIA Años 2007-2009

La presente memoria destaca como hechos significativos los siguientes.

• Congreso SABI 2007

El XVI Congreso Argentino de Bioingeniería y las V Jornadas de Ingeniería Clínica (SABI2007) se llevaron a cabo en la ciudad de San Juan durante los días 26 al 28 de Septiembre de 2007, organizado por la Regional Cuyo de la Sociedad Argentina de Bioingeniería y por la Facultad de Ingeniería de la Universidad Nacional de San Juan.

El congreso SABI2007 contó con la presencia de 524 asistentes de los países: Chile, Perú, Uruguay, Brasil, Colombia, Estados Unidos, España, Italia, Alemania y Argentina. El mismo fue declarado de interés provincial por el Gobierno de la Provincia de San Juan y patrocinado por la Agencia Nacional de Promoción Científica y Tecnológica (ANPCyT), el Consejo Nacional de Investigaciones Científicas y Técnicas (CONICET), the Institute of Electrical and Electronic Engineers (IEEE) Sección Argentina y por diferentes empresas locales y nacionales.

Se recibieron 186 contribuciones científicas, de las cuales fueron aceptados finalmente 152 trabajos. Se realizaron 7 conferencias plenarias a cargo de destacados investigadores extranjeros provenientes de Chile, España, Estados Unidos y nacionales.

En el marco de actividades de SABI2007, se llevó a cabo un concurso para premiar los mejores trabajos estudiantiles de grado y posgrado. Se presentaron 20 trabajos para el concurso de grado y 23 trabajos para el de posgrado.

Un día antes del congreso se llevó a cabo la Jornada de SABI Estudiantil. La misma contó con la participación de 214 estudiantes, provenientes de Entre Ríos, Mendoza y San Juan; docentes, egresados y público en general. Se expusieron 17 trabajos en diferentes sesiones y se realizó una Mesa Redonda, en la cual varios bioingenieros compartieron sus experiencias laborales en los distintos ámbitos externos a la universidad vinculadas al quehacer en la bioingeniería

• 1er Número especial de la Revista de SABI

Se realizo en el año 2007 el primer número especial de la Revista, en este caso fue sobre Ingeniería Clínica, además de la edición semestral, con una tirada de 500 ejemplares por número. El número especial fue coordinado por el Ing. Eduardo de Forteza, Presidente del Capítulo de Ingeniería Clínica de SABI.

• XII Jornadas Internacionales de Ingeniería Clínica y Tecnología Médica.

Durante los día 20 al 22 de Agosto del 2008 se llevaron a cabo estas jornadas en la ciudad de Paraná, Entre Ríos, organizadas por el Grupo de Estudios en Ingeniería Clínica de la Universidad Nacional de Entre Ríos (GEIC). Se contó con la participación de 206 profesionales y estudiantes de 7 países de Latinoamérica. En dicho marco se realizaron un total de 37 exposiciones orales y 20 presentaciones tipo poster y 3 Talleres de capacitación, estos últimos con cupo de asistentes completo. Estas Jornadas contaron con el auspicio de asociaciones como la American College of Clinical

Engineering (ACCE), la Asociación Peruana de Ingeniería Clínica (APIC), el Comité Argentino de Mantenimiento (CAM), la Cámara de Equipamiento Hospitalario de Fabricación Argentina (CAEHFA), universidades de América latina y empresas de equipamiento médico.

• Nuevo sitio web de SABI

La Pagina web de la sociedad paso a la administración de la Regional Rosario, siendo el Web master Sr. Mario Figge (mario@3tristestigres.com). Además se tomo el dominio www.sabi.org.ar

• 2do Número especial de la Revista de SABI

Se realizo en el año 2008 el segundo número especial de la Revista, en este caso fue sobre Biomecánica e Ingeniería en Rehabilitación, además de la edición semestral, con una tirada de 500 ejemplares por número. Este número especial fue coordinado por Mst. Bioing. Carolina Taberning, Mst. Bioing. Rubén Acevedo y el Dr. Bioing. Ariel Braidot.

• Reunión de la BioRed

De conformidad al acta acuerdo de Cooperación Académica, Científica y Tecnológica para las carreras de grado de Bioingeniería e Ingeniería Biomédica de las 6 Unidades Académicas del país, se reunieron sus representantes en la ciudad de Córdoba en el mes de julio del 2008. Se coordinaron las actividades del Subproyecto de Intercambio del Plan Plurianual 2008-2010 del Proyecto de Mejoramiento de la Enseñanza en Ingeniería (PROMEI II). Se plantearon estrategias en los sistemas de apoyo a la investigación, incumbencias profesionales, participación en organismos estatales como el ANMAT.

Dicha reunión contó con 2 invitados: El Director de Tecnológica Médica del ANMAT y el Presidente del Consejo de Profesionales de la provincia de Córdoba.

• 2 das. Jornadas Argentinas sobre Interfaces Cerebro Computadora JAICC 2009.

El Laboratorio de Ingeniería en Rehabilitación e Investigaciones Neuromusculares y Sensoriales (LIRINS) de la Facultad de Ingeniería de la UNER, en conjunto con la regional litoral de SABI, organizó las 2das Jornadas Argentinas sobre Interfaces Cerebro Computadora durante los días 23 y 24 de Abril del 2009 en Paraná, Entre Rios.

Se recibieron trabajos no solo de Argentina sino también de Chile, Colombia y México; lo cual le otorgó un carácter latinoamericano al evento. Se contó con invitados de reconocido prestigio, tanto nacionales como internacionales, provenientes de Suiza, Francia, México y Chile.

• Convenio con SADIO

En el mes de Agosto del 2009 se firmo un Convenio con la Sociedad Argentina de Informática SADIO, participando asimismo de una Mesa de Enlace con otras sociedades medicas como la AMA (Asociación Medica Argentina) y AMIA (Asociación Medica Informática Argentina) con el fin de coordinar actividades conjuntas.

• Congreso SABI 2009

En el Centro Cultural Bernardino Rivadavia (CCBR) de la Ciudad de Rosario, durante los días 14, 15 y 16 de Octubre de 2009 se llevó a cabo el XVII Congreso Argentino de Bioingeniería y las VI Jornadas de Ingeniería Clínica (SABI2009). El mismo fue organizado por la Regional Rosario de la Sociedad Argentina de Bioingeniería y el Polo Tecnológico de Rosario. Contó con el auspicio del Consejo Nacional de Investigaciones Científicas y Técnicas (CONICET), the Institute of Electrical and Electronic Engineers (IEEE) Sección Argentina, de la Cámara de Diputados del Gobierno de la provincia de Santa Fe, de la Universidad Abierta Interamericana, de Agencia Española Cooperación Internacional (AECID), organismos de control (ANMAT) y por diferentes empresas locales y nacionales.

Contó con la asistencia de 750 personas y se recibieron 203 contribuciones científicas, de las cuales fueron aceptados finalmente 183 trabajos. Se realizaron además conferencias plenarias a cargo de 10 destacados investigadores extranjeros provenientes de Uruguay, Chile, Brasil, Paraguay, Perú, Venezuela, Colombia, Cuba, España y EEUU.

Se premiaron los mejores trabajos estudiantiles en las categorías de grado y posgrado.

En forma simultánea se desarrolló la Jornada estudiantil SABI 2009, que contó con la presentación de 14 trabajos, de los cuales 13 fueron aceptados provenientes de las Universidades de Tucumán, San Juan y Entre Ríos. Estas Jornadas se desarrollaron con una gran participación de estudiantes de las distintas unidades académicas de las carreras de Bioingeniería e Ingeniería Biomédica del país.

• Creación del Capítulo Ingeniería en Rehabilitación

En oportunidad del último Congreso SABI 2009 y mediante resolución de la Asamblea General de Socios, se definió la creación del Capítulo Ingeniería en Rehabilitación, fundado por los profesionales e investigadores miembros de la Sociedad que desarrollan tareas en dicha temática. Los integrantes de este Capítulo discutieron acerca de los objetivos básicos, definieron su organización, modo de funcionamiento y designaron como Presidente al Mg. Ing. Juan Pablo Graffigna.

• Asamblea Ordinaria de SABI

El día 15 de Octubre de 2009, en el marco de las actividades del Congreso SABI 2009, se reunió la Asamblea Ordinaria de la Sociedad Argentina de Bioingeniería en el Centro Cultural Bernardino Rivadavia de la ciudad de Rosario. Entre los temas del orden del día, se trataron: Tratamiento y aprobación de la memoria, inventario, balance e informe de la Comisión Revisora de Cuentas de los períodos 01/08/07 al 31/07/08 y 01/08/08 al 31/07/09; y renovación de autoridades de la Sociedad.

La Asamblea contó con la presencia de la Veedora de Personería Jurídica Srta. Abogada Manso y 63 socios presentes, los cuales firman el Libro de Asistencia.

Se aprueban la memoria, el balance, e inventario y la Comisión Directiva saliente eleva una propuesta de Comisión Directiva; la cual fue aprobada. La conformación de la misma, es la siguiente:

PRESIDENTE:
Ana María Echenique (amechenique@gateme.unsj.edu.ar)
VICEPRESIDENTE:
Ariel Braidot (abraidot@bioingenieria.edu.ar)
SECRETARIA:
Virginia Ballarin (vballari@fi.mdp.edu.ar)
TESORERA:
Myriam Herrera (mherrera@herrera.unt.edu.ar)
1er VOCAL:
Juan Pablo Graffigna (jgraffig@gateme.unsj.edu.ar)
2do VOCAL:
Darío Rocha (dario@consultar.org)
1er VOCAL Suplente:
Juan Manuel Olivera (jolivera@herrera.unt.edu.ar)
2do VOCAL Suplente:
Diego Beltramone (dbeltramone@efn.uncor.edu)
Comisión Revisora de Cuentas:
Adriana Scandurra (scandu@fi.mdp.edu.ar)
y Ernesto Federico Treo (etreo@gateme.unsj.edu.ar)

Al finalizar la Asamblea se confirma a la Regional Mar del Plata, como organizadora del próximo Congreso SABI 2011.

En el Acto de cierre del Congreso de SABI 2009, la Nueva Comisión Directiva da a conocer sus propuestas:

• Generar estrategias que permita la participación activa como un referente nacional, facilitando la comunicación con los principales actores. Una estrategia es proponer la creación de un Consejo asesor a nivel de la Comisión Directiva de SABI con representantes de áreas estratégicas, tales como Secretaria de Ciencia y Tecnologías, Organismos Internacionales, CONICET, CONEAU, etc.
• Proponer y crear comisiones que trabajen en diferentes líneas con objetivos bien definidos, que permitan una mayor participación de todos los socios. Por ejemplo, una comisión de desarrollo tecnológico con la participación de las distintas regionales, a los efectos favorecer el establecimiento de proyectos de base tecnológica en bioingeniería.
• Proponer al Comité Administrativo de CORAL la posibilidad de designar a la Facultad de Ingeniería de la Universidad Nacional de Entre Ríos (UNER); como posible Institución anfitriona de la XXI Reunión de Coral en Agosto de 2010
• Organizar actividades programadas con las regionales con una propuesta inicial de una actividad rotativa por regional, sobre una temática definida y transmitirla al resto; por ej. por teleconferencia. Promover actividades (talleres, simposios, intercambio de investigadores entre los distintos centros, etc.) de los capítulos creados (Ingeniería Clínica e Ingeniería de Rehabilitación) y documentarlas.
• Constituir un comité Editorial de SABI, con un coordinador para la edición no solo de la revista, sino también de Libros con objetivos educativos y/o científicos de las distintas temáticas pertinentes a la Bioingeniería.

• Lograr la indexación de la Revista, generando el formato adecuado. Para lo cual es necesario un compromiso formal de todas las regionales para enviar trabajos que permitan aumentar el número de revistas anuales. Explorar la posibilidad de una edición paralela en formato digital, que permita mostrar el índice completo y la primera pagina del artículo.
• Propiciar actividades conjuntas con otras sociedades científicas, con las cuales se ha formalizado convenio (Ej. Sociedad Argentina de Informática e Investigación Operativa SADIO).
• Realizar reuniones virtuales mensuales de los miembros de la Comisión Directiva y delegados regionales.
• Generar una base de datos electrónica de socios, con un motor de búsqueda inteligente (temática de investigación, regional, etc).
• Evaluar la creación el capítulo de Informática Médica, propuesta recientemente acercada.

The College of Biomedical Engineering (CBME)

Biomedical Engineering in Australia

David Dewhurst, Australia's first biomedical engineer, attended the first ICMBE in Paris, in 1959, and his involvement and enthusiasm kick-started Australia's long and active involvement in the activities of the IFMBE. Australia joined the Federation in 1967, and David was an early member of the Administrative Council. In 1971, under his Presidency, Melbourne hosted the 9th ICMBE. Then in 1989, Melbourne hosted the 2nd IFMBE Pan-Pacific Conference, and in 2003, the World Congress was held in Sydney.

The development of biomedical engineering and, indeed, of health care as a whole in Australia, has been shaped by geography and demographics. Each of the country's six states and two territories has its own priorities, problems and challenges. Most Australians live in the relatively fertile south east and the eastern seaboard, with over half the population in the large and highly urbanised cities of Sydney and Melbourne. The vast centre, north and west of the country are sparsely populated, with often huge distances between towns and cities.

IFMBE Administrative Council Meeting, Sydney, 2003

In most states, the Societies for Medical and Biological Engineering (SMBEs) – autonomous state-based interest societies – organise many of the day-to-day BME activities such as talks and seminars, often in conjunction with other BME and medical physics organisations.

The College of Biomedical Engineers (CBME) is the country's major professional BME organisation, a college of Engineers Australia, which is the country's largest professional engineering association. Its Colleges implement its learned society function at the national level, and the CBME is involved in activities such as BME education and training and links with industry and academia, with specialist sub-committees covering Rehabilitation Engineering, Biomechanics of Impact Injury, and Clinical Engineering

Australia's IFMBE member society, the Australian Federation for Medical & Biological Engineering, brings together the SMBEs, the CBME and the Australian biomedical engineer members of the Australasian College for Physical & Engineering Sciences in Medicine. These organisations combine to run an annual scientific meeting nationally, often in conjunction with a medical physics meeting, and in most states, they organise technical lecture and training programs.

Australian public hospitals are run largely by the states and healthcare delivery and priorities vary considerably from state to state. Teaching hospitals in the more densely populated states generally have well established clinical engineering departments, while many hospitals in the less densely populated states rely on travelling technical support.

The overwhelming proportion of Australian biomedical engineers are clinical engineers employed by hospitals, government departments, rehabilitation centres, and universities. Some are involved in the development of medical devices and the development and transfer of medical services technology, while others are self-employed as consultants.

Newly elected IFMBE Officers and Administrative Council Members (L-R: J. Nagel, R. Magjarevic, S. Krishnan, M. Kikuchi, H. Terio, O. Barnea, K. Tsujioka, S. Laxminanayan)

Most BME research is located in universities, often in collaboration with clinical groups in teaching hospitals. Major research areas include prostheses, non-invasive measurement, biomechanics, and – most appropriately in a country of such vast distances and with an ageing population – rehabilitation, telemedicine, and home healthcare monitoring.

Undergraduate courses in Biomedical Engineering are available in ten Australian universities, several of these being 5 year combined double degrees in engineering and science. Some states also offer training programs for biomedical engineering technicians. Australia is a significant provider of biomedical education and training to tertiary students from Asia, and many of the Masters by coursework degrees being offered in biomedical engineering are primarily populated by students from Asia.

Australia's biomedical industry, though not large by world standards, and hit hard by globalisation, is very active. In addition to the two largest and internationally known BME firms, Cochlear (the Cochlear Ear Implant, digital hearing aids etc) and Resmed (products for the diagnosis and treatment of sleep disordered breathing), a second tier of smaller firms includes about one hundred producing specialist diagnostic and monitoring equipment and components.

In recent years, Australia's most important BME contribution to the region is the Medical Equipment Management in the Pacific program, which forms part of Australia's international aid program. The Project covers seven Pacific Island countries, with by far the largest program in Papua, New Guinea. The driving force behind the program is the need to provide better medical equipment and support for local hospitals in these countries.

Richard L.G. Kirsner
Hon Secretary, AFMBE
rkirsner@bigpond.com

Austrian Society for Biomedical Engineering OeGBMT

The Austrian Society for Biomedical Engineering OeGBMT was founded in 1975 at the Technical University of Graz by Stefan Schuy, the first Full Professor for "Biomedical Engineering" in Austria. From the very beginning the Society was dedicated to supporting the collaboration of technology and medicine in research, education and practical application. The number of members was growing continuously over the years to actually stabilize between 200 and 250, including senior-, honorary-, supporting and institutional members. OeGBMT represents Austria in the international umbrella organisations IFMBE and EAMBES and stands in close relation to the German speaking sister-organizations in Switzerland (SGBT) and Germany (DGBMT). Every third year the annual meeting is a jointly organized 3-country event in one of the three countries (2010 in Rostock, Germany).

Assignments and aims of OeGBMT have grown continuously with development of the field. Whereas initially electrical and mechanical technology, chemistry and medicine were linked in a multidisciplinary network, meanwhile biotechnology and informatics originated and gain growing importance. The field of education developed with similar dynamics and grew in diversity and width: it offers from selective biomedical engineering graduation on a high school level over a wide range of highly specialized curricula at various university and sub-university levels in public and private institutions to postgraduate courses. Consequently support of the upcoming generation of professionals has become one of our main efforts of our society. Our task force on "Pre- and Postgraduate Education", established in 2006, is focused on these developments at national, but also international level. To this end the society especially supports young scientists via dedicated competitions for diploma and PH.D. students, and by the annual Stefan-Schuy-Award for the publication by a young scientist. Not at least we offer a web-based job fair to support access to academic and industrial job market.

A further important focus of OeGBMT is the representation of the Austrian scientific community of "Biomedical Engineering" in relation to public bodies and authorities, e.g. in questions of research funding, education guidelines, medical product certification and ethical aspects.

The future of OeGBMT will without doubt require consequent and continuous efforts to follow the inherent dynamics of the social environment and associated developments of the field. With intensified networking on international level Biomedical Engineering will gain highly relevant and lasting importance for creation of added value in economy and improvements in the health care system. In this context we build on close and intensive co-operation with IFMBE. We want to take this occasion to congratulate the federation on her 50th anniversary and to thank for the longstanding fruitful collaboration over many years.

Ernst Hofer
(ÖGBMT, Past President)

Winfried Mayr
(ÖGBMT, President)

Please find detailed and continuously updated information on our society website http://www.oegbmt.at

IFMBE ProTem Working Group Meeting in Vienna, 2002.
(L-R: H. Hutten, J. Wojcicki, N. Saranammi, C. Roux, J. Nagel, D. Jaron)

Die Österreichische Gesellschaft für Biomedizinische Technik ÖGBMT

Die Österreichische Gesellschaft für Biomedizinische Technik ÖGBMT wurde 1975 an der Technischen Universität Graz von Stefan Schuy, dem ersten österreichischen Ordinarius für das Fach "Biomedizinische Technik", gegründet. Die Gesellschaft war von Beginn an der Förderung der Zusammenarbeit von Technik und Medizin in Forschung, Ausbildung und praktischer Umsetzung gewidmet. Die Zahl der Mitglieder ist über die Jahre kontinuierlich angestiegen und liegt in den letzten Jahren stabil zwischen 200 und 250 inklusive Senioren-, Ehren-, Fördernden und Institutionellen Mitgliedschaften. Die ÖGBMT vertritt die österreichischen Interessen in den internationalen Dachorganisationen IFMBE und EAMBES und pflegt intensive Kontakte mit den deutschsprachigen Schwestergesellschaften in der Schweiz (SGBT) und Deutschland (DGBMT). Alle 3 Jahre wird die Jahrestagung als gemeinsam organisierte 3-Ländertagung in einem der drei Länder abgehalten (2010 in Rostock, Deutschland).

Die Aufgaben und Ziele der ÖGBMT sind mit der Entwicklung des Wissensfeldes ständig gewachsen. War die multidisziplinäre Vernetzung zunächst auf die Fächer Elektrotechnik, Maschinenbau, Chemie und Medizin konzentriert, sind inzwischen Biotechnologie und Bioinformatik immer wichtiger geworden. Mit ähnlicher Dynamik hat die Ausbildungslandschaft an Vielfalt und Breite zugenommen: das Angebot reicht von punktuellen Fachausbildungsmöglichkeiten mit Fachmatura über eine breite Palette an Curricula auf verschiedenen Fachhochschulen und Universitäten im öffentlichen und privaten Sektor bis zu postgraduellen Lehrgängen. Dieser Entwicklung Rechnung tragend hat sich die Nachwuchsförderung zu einer der wichtigsten Aufgaben der Gesellschaft entwickelt: Die 2006 gegründete Arbeitsgruppe "Aus- und Weiterbildung in Biomedizinischer Technik" widmet sich dieser Entwicklung auf nationalem aber auch internationaler Ebene. Die Gesellschaft fördert gezielt den wissenschaftlichen Nachwuchs durch spezielle Wettbewerbe für Diplomanden und Dissertanten sowie durch die jährliche Ausschreibung und Vergabe des Stefan-Schuy-Preises für die beste Publikation eines Nachwuchsforschers. Schließlich bietet eine Jobbörse Hilfe beim Zugang zum akademischen und industriellen Arbeitsmarkt.

ÖGBMT sieht sich darüber hinaus als Vertreterin der österreichischen "Scientific Community" ihres Fachgebietes gegenüber nationalen und internationalen öffentlichen Stellen und Behörden, beispielsweise in Fragen der Forschungsförderung, Ausbildungsrichtlinien, Medizinproduktzulassungen oder ethischen Rahmenbedingungen.

Die Zukunft der ÖGBMT wird ohne Zweifel eine konsequente Weiterentwicklung nehmen entsprechend der inneren Dynamik der gesellschaftlichen Rahmenbedingungen und der damit verbundenen Weiterentwicklung des Fachgebietes. Mit zunehmender Vernetzung auf internationaler Ebene wird auch auf nationaler und regionaler Ebene die Biomedizinische Technik eine überdurchschnittlich nachhaltige Bedeutung in Wertschöpfung und Gesundheitswesen erlangen. In diesem Sinn setzen wir auf enge und intensive Zusammenarbeit mit der IFMBE, der wir an dieser Stelle besonders zum 50-jährigen Bestehen gratulieren und für die langjährige gute Zusammenarbeit danken wollen.

Ernst Hofer
(ÖGBMT, Past President)

Winfried Mayr
(ÖGBMT, President)

Nähere und laufend aktualisierte Informationen finden sich auf der Gesellschafts-Webseite http://www.oegbmt.at

EMBEC 2002, Vienna - Keynote speaker, Prof. Peter Schuster
with Profs. Keith Boddy and Dov Jaron in the audience at the Opening Ceremony.

The Belgian Society for Medical and Biological Engineering and Computing

Our society was founded in 1979, by August Delaunois, René Bourgain, Johan Colle, Marc Peeters, and Raoul Leclercq, with the purpose to organize scientific meetings and to publish the scientific results of these meetings. The founders were professors (physiology), engineers, and physicians, this still corresponds to the profile of our membership.

Since the beginning, the Society made great efforts to link up with the international community and this resulted in the representation of Belgium within the IFMBE.

Since the year 2000, the Belgian National Committee on Biomedical Engineering was recognized by the Royal Academy, which made it possible to organize our annual 'National Day on Biomedical Engineering', during which famous international speakers as well as members and research teams present their results.

Moreover, the Society organized 2 major European conferences: the North Sea Conference in 1990, attracting over 600 participants, and the BEC2008 Conference with 950 participants, mostly young researchers.

Participants on European IFMBE BME Conference 2008,
Antwerpen, Belgium

(L-R: S. Krishnan, C. Nyssen, M. Nyssen, D. Slaaf, H. Voigt, M. Secca)

De ´Belgian Society for Medical and Biological Engineering and Computing´

Onze vereniging werd in 1979 opgericht door August Delaunois, René Bourgain, Johan Colle, Marc Peeters en Raoul Leclercq, met als doel de wetenschappelijke bijeenkomsten te organiseren en de resultaten hiervan bekend te maken. De stichters waren hoogleraren (fysiologie), ingenieur en ziekenhuisarts, wat nog steeds het meest voorkomend profiel is van onze leden.

Sedert haar beginperiode zocht de vereniging aansluiting bij internationale verenigingen en werd zo de vertegenwoordiger van België in de IFMBE.

Sedert 2000 werd het Nationaal Comité Biomedical Engineering erkend door de Koninklijke Academie en zo werd het mogelijk een jaarlijkse ´National Day Biomedical Engineering´ in te richten, waarop zowel uitgenodigde buitenlandse sprekers als leden en onderzoeksteams hun resultaten voorstellen.

De vereniging organiseerde bovendien 2 grote Europese conferenties: de ´North Sea Conference´ in 1990 met meer dan 600 deelnemers en de ´MBEC2008 conference´ met 950 deelnemers.

Actueel hecht de vereniging grote aandacht aan verbreding, via de afgestudeerden aan de nieuwe master-programma´s in Biomedical Engineering en via contacten met de beroepswereld. Daarnaast worden internationale contacten goed onderhouden via afvaardiging in alle belangwekkende verenigingen, in verband met biomedical engineering.

Voorzitter (ook van Nationaal Comité): Prof. Dr. Ir. Pascal Verdonck
Secretaris (ook van Nationaal Comité): Prof. Dr. Ir. Marc Nyssen
Schatbewaarder: Ir. Dirk De Wachter

Leden Nationaal Comité:

Prof. Dr. Ir Jos Vander Sloten	Prof. Dr. Philippe Kolh
Prof. Dr. Jean Delbeke	Prof. Dr. Ir. Jan Cornelis
Prof. Dr. Jean Brimioulle	Prof. Dr. Ir. Alain Delchambre
Prof. Dr. Ignace Lemahieu	Prof. Dr. Floris Wuyts
Prof. Dr. Philippe Lefevre	Prof. Dr. Denis Flandre
Prof. Dr. Ir. Georges Van der Perre	

Ere-leden:
Prof. Dr. René Bourgain, Prof. Dr. Claude Veraart.

Secretariaat: c/o Prof. Marc Nyssen
Medical Informatics Dept. VUB
Laarbeeklaan 103
B-1090 Jette
Belgium

Opening Ceremony at European IFMBE BME Conference 2008,
Antwerpen, Belgium

Biomedical Engineering Brazilian Society (SBEB)

The Biomedical Engineering Brazilian Society-SBEB was founded in December 18, 1975.

SBEB gives a fundamental contribution towards the academic/scientific developing Biomedical Engineering in Brazil. Since their foundation SBEB organized several conferences and workshops around Brazil and edited scientific publications in Biomedical Engineering. They also interacts with government agencies and professional associations.

SBEB contribute for the exchanges of experiences and the establishment of policies for public health and technology, stimulating the creation and development of undergraduate and graduated schools of Biomedical Engineering as well as research groups. This society brings together hundreds of researchers working in the areas of Bio-engineering, Medical Engineering, Clinical Engineering, and others.

In Latin America SBEB is affiliated to CORAL-Regional Council for Latin America in Biomedical Engineering. SBEB is also affiliated to IFMBE-International Federation of Medical and Biological Engineering, which is the maximum representative of Biomedical Engineering in the world.

In order to promote the development of Biomedical Engineering, as well as to encourage scientific and technical exchanges among its members, SBEB is reporting the Biomedical Engineering activities in Brazil through the RBEB - Revista Brasileira de Engenharia Biomédica – Brazilian Journal of Biomedical Engineering, also organize at each 2 years the Brazilian Congress of Biomedical Engineering that in 2010 will be at 22ª edition.

CLAEB 2004 Opening Ceremony, Joao Pessoa, Brazil

Sociedade Brasileira de Engenharia Biomédica-SBEB

SBEB

A Sociedade Brasileira de Engenharia Biomédica-SBEB foi fundada em 18 de dezembro de 1975

A SBEB contribui de maneira fundamental para o desenvolvimento acadêmico/científico da Engenharia Biomédica no Brasil. Desde sua fundação a SBEB organiza várias conferências e workshops no Brasil editando publicações científicas na área. Ela também interage com diferentes agências governamentais e associações de profissionais.

A SBEB contribui para a troca de experiências e também para estabelecer políticas na área de saúde pública e tecnologia, estimulando a criação e o desenvolvimento de cursos de graduação e pós graduação em Engenharia Biomédica, bem como de grupos de pesquisa. Esta sociedade congrega centenas de pesquisadores trabalhando em Bioengenharia, Engenharia Médica, Engenharia Clínica entre outras áreas.

Em nível de América Latina a SBEB é afiliada ao CORAL-Conselho Regional para América Latina de Engenharia Biomédica e em nível mundial é afiliada à IFMBE-International Federation of Medical and Biological Engineering, órgão com máxima representatividade em Engenharia Biomédica no mundo.

De maneira a promover o desenvolvimento da Engenharia Biomédica, bem como encorajar as trocas tecno-científicas entre seus membros, a SBEB publica as atividades de Engenharia Biomédica no Brasil, através da Revista Brasileira de Engenharia Biomédica-RBEB e também organiza a cada 2 anos o Congresso Brasileiro de Engenharia Biomédica que em 2010 realizará sua 22ª edição.

The Bulgarian Society of Biomedical Physics and Engineering (BSBPE)

Historical Survey of the BSBPE and BME in Bulgaria

In 1869, the Bulgarian Literary Society was founded and later developed into the Bulgarian Academy of Sciences (BAS). Some years later the Bulgarian engineers and architects formed their own organization. The first biomedical engineer in Bulgaria was S. Karadimov who introduced X-ray and electrical therapy instrumentation in the country before the Second World War and established a service and maintenance facility. These activities were further developed by the X-ray technicians S. Stefanov and V. Stefanov. Since 1956, the young engineers K. Popov and E. Milev contributed to the maintenance of the more complex medical equipment. Under the supervision of S. Karadimov the research group: K. Marinchev, D. Karadimov, I. Stamboliev and I. Daskalov started local production of electrotherapy devices. Several years later L. Jovev and K. Popov, assisted by I. Dotsinsky, established the Institute of Medical Engineering whose main activity was the development and production control of medical instrumentation.

Later the Department of Biomedical Engineering, a research and clinical engineering unit, within the Medical Academy (today the Medical University in Sofia) was built up, headed by Prof. I. Daskalov. In the 1990s, this Department has been co-opted to the BAS as the Center of Biomedical Engineering. I.Stamboliev initiated the education of medical engineers in 1974. A specialization in medical electronics was set up within the Department of Electronics Engineering at the Technical University of Sofia. In the last few years, educational activities in biomedical engineering were undertaken at the Technical Universities in Varna, Plovdiv (a Division of the Technical University in Sofia) and Gabrovo as well. Considerable improvement was achieved due to the EC supported Project "TEMPUS" – an international postgraduate course. Despite some decline in the interest in engineering sciences after the political and economic changes in 1989, about 45 students graduated in medical electronics.

From 1962 to 1971, the "biomedical engineers" were members of the Medical Electronics Section of the Bulgarian Physiological Society. In 1971, they merged with the medical physicists to form the new Bulgarian Society of Biomedical Physics and Engineering (BSBPE). Among the founders of the Society were D. Karadimov, Stamboliev, and Prof. I. Daskalov. At the World Congress 2003 in Sydney, Prof. I. Daskalov (1933-2004) was elected as a regular member of the International Academy of Biomedical Engineering.

Some of the principal concerns of the BSBPE at present are: organization of national conferences, distribution of information on scientific research and development from the country and abroad as well as practical implementations, expertise, revision of new books, approbation of dissertations or publications, participation in the education and postgraduate training of medical physicists and engineers from Bulgaria and abroad, consultations to governmental and non-governmental institutions on major problems of medical physics and engineering such as acquisition of special instrumentation and distribution of clinical physics and engineering services, and participation in international projects within the EC 6th and 7th Frame Programs, PHARE, etc. Through the years the activities varied and have been carried out with different intensity. Since 1972, the Society holds its regular scientific National Conferences every four years. Scientists from all Central and Eastern European countries, France, Germany, England, India, the USA, Canada and Australia actively participated in the ten conferences by presenting papers and joining the discussions. The scope of the conferences included WHO-seminars, IOMP- and EFOMP-supported meetings and IAEA-workshops which were organized in order to follow actual problems and novel trends in special fields: modern procedures and instrumentation, quality assurance, harmonization of standards and legislation, education and training among others.

The Bulgarian Society of Biomedical Physics and Engineering (BSBPE) is a collective member of the Union of Scientists in Bulgaria, the Union of Bulgarian Medical Societies, the International Federation of Medical and Biological Engineering (IFMBE), the European Association of Medical Biological Engineering and Sciences (EAMBES), the International Organization of Medical Physics (IOMP), and the European Federation of the Organizations for Medical Physics (EFOMP). A good collaboration is realized amongst all of them. The BSBPE is maintaining good contacts with identical organizations in Germany, Czech Republic, Poland, Hungary, Greece, Turkey, etc., as well as the WHO, the IAEA, and CERN among others.

Scientific research and development work in BME is mainly concentrated in several institutions in Sofia: the Centre of Biomedical Engineering with Prof. I. Daskalov, the bioengineering groups at the Institute of Physiology and the Institute of Biophysics at the Bulgarian Academy of Sciences, the Department of Medical Electronics at the Technical University in Sofia, the Clinical Engineering Department at the Medical University, and some years ago at the Military Medical Academy. More than 500 papers on different topics of BME have been published in international scientific journals and books. Clinical Engineering has a prolonged tradition in Bulgaria. Through the years about 800 engineers and technicians have been occupied in this field. Some of them are engaged in separate service offices or as representatives of foreign manufacturers, others accomplish their duties as clinical engineers in large hospitals throughout the country. The majority of them graduated in medical electronics at the Technical Universities in Sofia and Varna. The medical engineering industry used to include assembly of X-ray units manufactured by foreign producers in the locally developed X-ray units and electrotherapy devices, ECG's and standard blood pressure measurement devices, surgical instruments, prosthetics and others.

Commission for the Advancement in Healthcare Technology Management in Asia (CAHTMA)

The medical device industry in Malaysia has advanced in leaps and bounds during recent years encompassing various disciplines from clinical practices, maintenance and healthcare organizations, device manufacturers, and distributors as well as patients. In responding to rapid regional compliance to medical device regulations, industry stakeholders have increasingly tailored their organizational deliverables towards safety, in particular, patient safety resulting in improved patient care throughout. Commission for the Advancement in Healthcare Technology Management in Asia (CAHTMA) was established in part to ensure the interest of patient safety is held paramount.

Over the past several years, CAHTMA has developed and redeveloped a comprehensive clinical engineering certification program for technicians, engineers, and other technical personnel covering the medical industry. A number of clinical engineering personnel have been certified and many more will choose certification in the coming years in view of the increased regulation of the medical industry in Asia. CAHTMA has also received interest of its certification program and services from countries in Asia and the Middle-East. The activities to fulfill these requests have been undertaken over the past years and will continue to be undertaken in the coming years.

Apart from certification, CAHTMA has been active in clinical engineering activities in the region. Amongst activities undertaken by CAHTMA were as follows:

- Carried out assessment and training on medical equipment in Vietnam
- CAHTMA was part of the faculty on ORBIS's effort on clinical engineering in Bangladesh
- In collaboration with Healthtronics, a clinical engineering company in Malaysia, CAHTMA helped to train clinical engineering managers, engineers, and technicians from Vietnam, Maldives, and Iraq.
- CAHTMA coordinated clinical engineering workshop/seminar on electrical safety for stakeholders in Malaysia.

Although relatively new in the region, certification is expected to gain strength within the next two years. In a proactive move, CAHTMA have been developing certification modules for equipment users, an area often overlooked by device manufacturers and hospital management alike.

Canadian Medical and Biological Engineering Society (CMBES)

The Canadian Medical and Biological Engineering Society is Canada's principal society for engineering in medicine and biology. It is affiliated with the International Federation for Medical and Biological Engineering (IFMBE).

The Society's aims are twofold: scientific and educational: directed toward the advancement of the theory and practice of medical device technology; and professional: directed toward the advancement of all individuals in Canada who are engaged in interdisciplinary work involving engineering, the life sciences and medicine.

The CMBES organizes national medical and biological engineering conferences in various cities across Canada. In addition, the CMBES conference sponsors seminars, which include peer-reviews, technical paper presentations, exhibits of current medical devices, a continuing education program, workshops, symposia, and networking opportunities.

The CMBES Clinical Engineering Standards of Practice was published in 1998. These guidelines outline criteria for health care institutions on the management of medical devices, promote the professional development of its members, and outline the education and certification requirements for clinical engineers and biomedical engineering technologists and technicians.

The Society actively represents the interests of biomedical and clinical engineering to various government agencies by providing advice, submitting briefs, and reviewing documents. These agencies include:

- Canadian Council for Health and Services
- Health Canada
- Medical Devices Regulatory Committee
- Ministries of Health
- Canadian Hospital Association
- CSA Health Technology Program

The CMBES facilitates communication among its membership and other agencies by regularly publishing a newsletter and the proceedings of all conferences and seminars. A career booklet is also available for guidance counsellors and employment centers. The Society also maintains an information document pertaining to biomedical engineering education available in Canada.

La société canadienne de genie biomédical

La société canadienne de génie biomédical est la principale organisation canadienne pour l'ingénierie dans les domaines de la biologie et de la médecine. Elle est affiliée avec la Fédération internationale d'ingénierie en médecine et en biologie (IFMBE).

La Société a un double objectif : scientifique et éducatif : orienté sur l'avancement de la théorie et de la pratique de la technologie de l'équipement médical ; et professionnel : orienté sur l'avancement de tous les individus au Canada qui sont engagés dans le travail interdisciplinaire impliquant la technologie, les sciences de la vie et la médecine.

Au niveau national, la SCGB organise des conférences en technologies biologique et médicale dans diverses villes à travers le Canada. En outre, la conférence de la SCGB commandite des conférences qui incluent l'évaluation par des pairs, la présentation d'exposé technique, l'exposition des appareils médicaux courants, un programme de formation continue, des ateliers, des colloques, et des occasions pour tisser des liens professionnels.

Les normes de bonnes pratiques de la SCGB ont été éditées en 1998. Ces directives décrivent des critères à l'endroit des établissements de soins de santé sur la gestion des appareils médicaux, favorisent le développement professionnel de ses membres, et décrivent les conditions de formation et de certification pour les ingénieurs cliniques, les technologues et les techniciens biomédicaux.

La Société représente activement les intérêts de la technologie biomédicale et clinique à divers organismes gouvernementaux en fournissant des conseils, en soumettant des dossiers, et en examinant des documents. Ces agences incluent:

- Conseil canadien d'agrément des services de santé
- Santé Canada
- Comité de réglementation d'appareils médicaux
- Ministères de la santé
- Association canadienne des soins de santé
- Programme de technologie de santé de CSA

La SCGB facilite la communication entre ses membres et d'autres agences en éditant régulièrement un bulletin et des comptes rendus de conférences et séminaires. Un livret de carrière est également disponible pour des conseillers d'orientation et des centres d'emploi. La Société maintient également un document d'information concernant la formation en technologie biomédicale disponible au Canada.

The Chinese Society of Biomedical Engineering (CSBME)

The Chinese Society of Biomedical Engineering (CSBME), established in 1980, is a voluntary, national, and academic organization registered in accordance with the law for professional personnel involved in academic activities of engineering sciences in medicine and biology. It is directly affiliated with the China Association for Science and Technology; and is a major driving force for the development of biomedical engineering (BME) in China.

The major goals of the CSBME are: to provide a scientific platform for biomedical engineers and scientists to exchange new ideas and achievements in the BME field; to promote the development of BME in China by organizing national and international BME scientific conferences, and publishing its academic journals; to establish links between academic industrial researchers as well as commercial vendors; to provide advisory services for China's major decision-making in related fields; and to introduce modern technology for improving health care in China.

Participants of APCBME 2008, Beijing, China

Professor Jiasi Huang, one of the founders of the CSBME and the late president of Chinese Academy of Medical Sciences, was the president of the 1st Council of the CSBME. Since then, the Council has been elected every four years. Now it is the 7th Council which consists of 68 council members including 28 standing members. The current president is Professor Yubo Fan, Dean of School of Biological Science and Medical Engineering, Beihang University. The CSBME now has over 3500 registered members and 22 affiliated groups including 17 regional BME organizations.

The CSBME presently has 20 scientific sub-societies. These include: Artificial Organs, Biomaterials, Biomechanics, Biomedical Measurement and Instrumentation,

Biological Information and Cybernetics, Medical Physics, Medical Ultrasonic Engineering, Heart Rhythm, Bioelectromagnetics, Biomedical Sensor Technology, Clinical Medical Engineering, Traditional Chinese Medicine Engineering, Hemotherapy Engineering, Extracorporeal Circulation, Medical Equipment, Tissue Engineering and Regenerative Medicine, Stem Cell Engineering Technology, Tumor Targeted Therapy Technology, Digitalized Medicine and Medical Information Technology, and Intervention Medicine Engineering. In the Council, there are 11 Working Committees covering Organization, Academic activity, Editing, International-affair Liaison, Science Popularize and Vocational Training, Education, Industry Promotion, Clinic, Youth, Honorary Experts, and Awards.

The CSBME holds its General Assembly and National Academic Congress on BME every 4 years, and its sub-societies hold academic activities in their own fields every year. By the end of 2007, the CSBME had held 270 different kinds of academic activities and organized or participated in over 90 international conferences, during which over 20,000 papers were presented. In addition, from 2003 to 2008, it has held 18 various training programs in total with about 3,000 participants. There are four academic journals published by the CSBME: *Chinese Journal of Biomedical Engineering* (Chinese and English edition), *Chinese Journal of Hemorheology*, and *Chinese Journal of Cardiac Pacing and Electrophysiology*.

The CSBME and its Sub-society on Medical Physics (CSMP) have become members of the International Federation for Medical and Biological Engineering (IFMBE) and the International Organization of Medical Physics (IOMP), respectively since 1986. Also, some members of the CSBME are the personal members of International Society of Artificial Organs (ISAO).

The CSBME will host the 12th World Congress on Medical Physics and Biomedical Engineering (WC2012) in Beijing in 2012. The CSBME warmly welcomes your participation in the WC2012.

Depei Liu, PhD

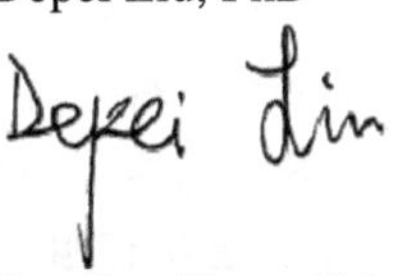

Past President, Chinese Society of Biomedical Engineering
Administrative Council Member, IFMBE
President, Chinese Academy of Medical Sciences & Peking Union Medical College

中国生物医学工程学会介紹

中国生物医学工程学会成立于1980年，是從事生物医学工程学科活動的科学技術工作者自願結成并依法登記成立的全国性学術團体，是中国科協的組成部分，是發展中国生物医学工程事業的主要社会力量。学会的主要任務是：爲生物医学工程領域的科技工作者提供学術交流平臺；通過組織国際、国内学術会議，出版学術刊物，推動中国的生物医学工程学科的發展；爲科技与産業的結合架設橋梁；爲国家相關領域重大決策提供咨詢服務；促進現代医療衛生技術和概念在中国的普及。

学会第一屆理事会由学会創始人之一、中国医科院已故院長黄家駟教授擔任理事長，此後理事会每四年改選一次，现爲第七屆，理事会规模68人，其中常務理事28人。现任理事長由北京航空航天大学生物与医学工程学院院長樊瑜波教授擔任。学会现有注册個人会員3500餘人，團体会員22個，其中包括17個地區性生物医学工程学術組織。

中国生物医学工程学会现有20個專業分会，領域涉及：人工器官、生物材料、生物力学、生物医学測量、生物信息与控制、医学物理、医学超聲工程、心律、生物電磁学、生物医学傳感技術、臨床医学工程、中医藥工程、血療工程、体外循環、衛生裝備、組織工程与再生医学、幹細胞工程技術、腫瘤靶向技術、数字医療及医療信息化、介入医学工程。学会下設11個工作委員会，工作涉及：組織、学術、編輯、外事、科普与職業培训、教育、産業促進、臨床、青年、老專家、獎励。

学会每四年舉行一次会員代表大会和全国生物医学工程学術会議，各專業分会每年進行專題学術交流。截至2007年底，学会共舉辦過各類国内学術交流活動270次，組織或参与国際学術会議90餘次，共交流論文2萬餘篇。此外，2003年至2008年期間共舉辦各類培训項目18項，培训約3000人次。学会主辦有四個学術期刊：中国生物医学工程学報，中国生物医学工程学報（英文版），中国心臟起傳与心電生理雜志，和中国血液流變学雜志。

中国生物医学工程学会于1986年加入国際医学生物工程聯合会（IFMBE）成爲国家会員；中国生物医学工程学会医学物理分会亦于1986年成爲国際医学物理学術組織（IOMP）成員。中国生物医学工程学会還有許多会員以個人身份参加了国際人工器官学会、国際生物材料学会等国際組織。

学会將于2012年在北京承辦国際医学物理和工程科学聯合会(IUPESM)每三年召開一次的世界学術大会（WC2012）。中国生物医学工程学会盛情邀請全世界的科学家参加WC2012，爲推動全世界生物医学工程事業的發展作出貢獻。

劉德培，工程院院士
中国生物医学工程学会前任主席
国際医学与生物工程聯合会執行委員会委員
中国医学科学院院長，中国協和医科大学校長

The Columbian Association of Biomedical Engineering and Medical Electronics (ABION)*

ABIOIN

The Colombian Association of Bioengineering and Medical Electronics (ABIOIN), was founded in 1993, in Bucaramanga, Santander, Colombia. Later, regional chapters were established in Antioquia (state), in the cities of the coffee axis, in the Caribbean region at Barranquilla, in Cundinamarca, and in the Valley of the Cauca. From this time it began to organize seminars, technological meetings, and symposia on specific topics of bioengineering. In 2003, ABIOIN sponsored the first Colombian Congress of Bioengineering and Biomedical Engineering in Medellin, Colombia, with 510 participants and 100 investigative projects.

With the purpose of positioning bioengineering near national government interests, ABIOIN organized the second Congress in the National Capitol, where the Senate of Colombia carries out its legislative actions. With an attendance of 650 participants, 200 works groups representing universities, companies, and Institutes related to Bioengineering were presented. In June 2008, the Association organized the third Congress in Pereira, Colombia. The attendance of 260 people and the presentation of 100 projects of the Bioengineering in Colombia reaffirmed the importance of taking Bioengineering in Colombia to all regions of the country. Other activities like varied Bioengineering symposia have been developed 2006 to 2008, organized by universities in Bogotá, Medellin, Cali, Barranquilla, Bucaramanga; among them were those organized by the universities of the Andes, of Javeriana, and of the Bogotá District. Also, at the Latin American level, the UAO organized the seminar for Latin American Clinical Engineering in September, 2008, in Cali, Colombia; the first congress of biomaterials, organized by the University of Antioquia, in September, 2008, in Medellin; the IV symposium of Bioengineering organized by the University of the Bogotá District. There were special conferences on nanotechnologies and other current topics that are related to bioengineering in different Colombian cities.

Every two years, research groups develop and prepare the work presented at the congresses along the various lines of the field of the bioengineering and biomedical engineering. The number of research papers in bioengineering is growing every period as is the number of scientific articles. About 300 scientific articles, on average, are presented for each event; a minimum of about 100 projects are selected to be presented in the congresses in different forms. These works become scientific publications originating from different research groups at the universities in Colombia, relative to bioinstrumentation, acquisition, processing, and control of the biomedical signals and images, clinical engineering, rehabilitation engineering, biomedical

* Isnardo Torres, National President of Colombian Association of Bioengineering (ABIOIN).

optics, biomaterials, nanotechnologies, biotechnology, and telemedicine. The most important universities regarding these investigations are[1]: UPB, EIA, UdeA, UIS, UN, UMB, UAN, UA, UJ, UD, UV, UAO, UAC, EAFIT, UC, and UTP, and others no less important.

The year 2008 marked the 50th anniversary of the pacemaker in Colombia; a time when Dr. Jorge Reynolds worked with the first external pacemaker installed in a patient until its current project of the nano-pacemaker. Other investigators as John Bustamante, Gabriel Mejia, Rafael Espinosa, Mauricio Wilches, and Martha Zequera have been working in the neurosciences, biotechnology, biomechanics, and a new group of young people who are making bioengineering one of the most attractive fields in academic, industry, and health.

There are several Institutions help to establish new companies in relation to biomedical engineering, such as IDEAR (Antioquia), and several business incubators in the main cities of Colombia. These companies develop specific skills for the development of new biomedical engineering companies in Colombia. The educational sector has a wide array of offerings from the undergraduate level to the graduate degree and specific extension short courses in the different bioengineering fields. Universities like UMB, UAN, UAO, USC, UN, UA, UD, UdeA, and EIA already have different degrees from the undergraduate to postgraduate degree. Other, short-term graduate courses, held in different specific areas such as clinical engineering, bioinformatics, and medical metrology, are growing in the different cities of Colombia. The reason is due to the high demand for quality services and accreditation that require medical services from hospitals and clinics. ABIOIN is associated with INVIMA[2] and other companies that are related to this area in Colombia. There is a favorable climate for the future of bioengineering in Colombia.

[1] EIA School of Engineering of Antioquia; UPB "Pontifical University Bolivarian; UdeA University of Antioquia; UIS Industrial University of Santander; UMB University Manuela Beltrán; UAN University Antonio Nariño; UD University Distrital Francisco José of Caldas; UN National University of Colombia; UV University of the Valley; UA University of the Andes; UJ University Javeriana; UAO Autonomous University of West; UTP Technological University of Pereira, US University of the Sabana; UC University of the Cauca.

[2] INVIMA Institute of surveillance and control of medications.

Las Actividades del Bioingenieria en Colombia*

La Asociación Colombiana de Bioingeniería y Electrónica Médica (ABIOIN) nació en 1993 en Bucaramanga, Santander, Colombia. Posteriormente se establecieron capítulos regionales en Antioquia, en las ciudades del eje cafetero, en la región Caribe, en Cundinamarca y Valle del Cauca. Desde esta época comenzó a organizar seminarios, rondas tecnológicas y simposios en temas específicos de Bioingeniería. En el año 2003 ABIOIN realizó su primer congreso Colombiano de Bioingeniería e Ingeniería Biomédica en Medellín, Colombia con la participación de 510 personas y 100 proyectos de investigación. Con el fin de posicionar la Bioingeniería dentro de los intereses gubernamentales se organizó el segundo Congreso en el Capitolio Nacional, lugar donde el Senado de Colombia realiza sus secciones del legislativo. Con una asistencia de 650 participantes se presentaron 200 trabajos de diferentes grupos de investigación de las universidades, empresas e Institutos relacionados con la bioingeniería. En Junio del 2008 se realizó el tercer Congreso en Pereira, Colombia. La asistencia de 260 personas y la presentación de 100 proyectos de la bioingeniería en Colombia reafirmo la importancia que esta tomando la bioingeniería en Colombia en todas las regiones del país. Se han desarrollados otras actividades como variados simposios de bioingeniería en el 2006 y 2008 organizados por universidades de Bogotá, Medellín, Cali, Barranquilla, Bucaramanga; entre ellos figuran los organizados por las universidades de los Andes, Javeriana, Distrital en Bogotá. También a nivel latinoamericano se ha organizado el seminario Latinoamericano de la ingeniería Clínica en septiembre del 2008 en Cali Colombia; el primer congreso de biomateriales organizado por la universidad de Antioquia en septiembre de 2008 en Medellín, IV simposio de Bioingeniería organizado por la Universidad Distrital en Bogotá. Hubo conferencias especialmente sobre nanotecnologías y otras temáticas de actualidad que tienen relación con la bioingeniería en las ciudades diferentes de colombiano.

Cada dos años los grupos de la investigación desarrollan y preparan los trabajos que presentan en los congresos en las diversas líneas del campo de la bioingeniería y la ingeniería Biomédica. El número de trabajos de investigación en bioingeniería está creciendo cada período y también el número o los artículos científicos. Sobre los 300 artículos científicos que en promedio presentan para cada evento se seleccionan mínimo unos 100 proyectos para ser presentados en los congresos en diferentes modalidades. Estos trabajos convertido en publicaciones científicas se originan de grupos diferentes de Investigación en las Universidades de Colombia relacionados con la Bioinstrumentación, la adquisición, procesamiento y control de la señales e imágenes biomédicas , la ingeniería clínica, la ingeniería de rehabilitación, la óptica

* Isnardo Torres, National President of Colombian Association of Bioengineering (ABIOIN).

biomédica, los biomateriales, la nanotecnología, la biotecnología y la telemedicina. Las universidades más importantes respecto a estas investigaciones son[1]: UPB, EIA, UdeA, UIS, UN, UMB, UAN, UA, UJ, UD, UV, UAO, UAC, EAFIT, US, UC y UTP entre otras no menos importantes

El año 2008 se celebró 50 años de aniversario del marcapaso en Colombia, tiempo dónde Dr. Jorge Reynolds han estado trabajando con los primeros marcapasos externos instalados a una persona hasta sus proyecto actual del nano marcapaso. Otros investigadores como John Bustamante, Gabriel Mejía, Rafael Espinosa, Martha Zequera han estado trabajando en el neurociencias, la biotecnología, la biomecánica y un nuevo grupo de personas jóvenes que están haciendo de la bioingeniería uno de los campos más atractivo a nivel académico, industrial y de la salud.

Existen varias Instituciones que ayudan a que se funden nuevas compañías de ingeniería biomédica como IDEAR(Antioquia) e diversas incubadoras de empresas en las ciudades principales de Colombia. Estás empresas desarrollan competencias específicas para el fomento de nuevas compañías de Ingeniería Biomédica en Colombia. La oferta de sector de educación es amplia desde el nivel de pregrado al postgrado y cursos cortos específicos de extensión en los diferentes campos de la bioingeniería. Las universidades como UMB, UAN, UAO, USC. UN, UA, UD, UdeA, EIA ya tienen títulos diferentes desde el pregrado al postgrado. Otros curso más cortos denominados diplomados se realizan en diferentes en las áreas específicas como ingeniería Clínica, bio informática, la metrología médica están creciendo en las ciudades diferentes de Colombia. La razón obedece a la alta demanda de servicios de calidad y acreditación que requieren los servicios Médicos de los hospitales y clínicas. ABIOIN se asocia a INVIMA[2] y otras compañías que tienen la relación a esta área en Colombia. Hay clima bueno para el futuro de bioingeniería en Colombia.

[1] EIA Escuela de Ingeniería de Antioquia; UPB Universidad Pontificia Bolivariana; UdeA Universidad de Antioquia; UIS Universidad Industrial de Santander; UMB Universidad Manuela Beltrán; UAN Universidad Antonio Nariño; UD Universidad Distrital Francisco José de Caldas; UN Universidad Nacional de Colombia ; UV Universidad del Valle; UA Universidad de los Andes; UJ universidad Javeriana; UAO Universidad Autónoma de Occidente; UTP Universidad Tecnológica de Pereira, US Universidad de la Sabana; UC Universidad del Cauca; USC Universidad Santiago de Cali.

[2] INVIMA Instituto de vigilancia y control of medicamentos en Colombia.

Croatian Medical and Biological Engineering Society (CROMBES)

The Croatian Medical and Biological Engineering Society (CROMBES) has more than a 100 members - scientists and experts involved in different scientific fields of biomedical engineering. CROMBES has been, and we hope will continue to be, the platform for interdisciplinary cooperation in biomedical engineering in Croatia and internationally. Members of CROMBES participate in various Croatian and international research projects and programs. Among these different projects, we especially emphasise the 5 year long scientific research project "Non invasive measurements and methods in biomedicine" endorsed by the Ministry of Science, Education, and Sports of the Republic of Croatia. The project is based on modern technologies and has the potential not just for further research and development but for the implementation of achievements in clinical practice and by industry.

Education in the field of biomedical engineering started in Croatia in the early 1970s, as a part of biomedical electronics teaching at the University of Zagreb Faculty of Electrical Engineering and Computing. Today, biomedical engineering is taught at several constituent institutions of the University of Zagreb as well as at other university centres in Croatia. Within higher education reform in Croatia, we plan to establish a biomedical engineering course at a master's level in such a way that teaching of different courses is performed at different constituent university institutions and also at other universities from the region. One of our objectives is also to increase the mobility of teachers and students. These efforts are also supported by the international Tempus project.

Prof. Stanko Tonković is awarded
Honorary Life President title

In the previous decade, CROMBES has significantly improved international cooperation with foreign universities through research projects. We have also organized several international scientific meetings like: 8th International IMEKO Conference on Measurement in Clinical Medicine, Dubrovnik 1998; 9th Mediterranean Conference on Medical and Biological Engineering and Computing-MEDICON 2001, Pula 2001; 1st International Summer School: Applications of ICT in Biomedicine, Dubrovnik 2002; and in cooperation with medical physicists 2nd Austrian, Italian, Slovenian and Croatian Medical Physics Meeting-AISCMP, Opatija 2006. CROMBES will also actively participate in the organization of 17th International Conference on Biomagnetism, Dubrovnik, spring 2010.

CROMBES is active in promoting clinical engineering and in cooperation with IFMBE it has been working on projects for registering and certifying clinical engineers as well as on preparing a program for their continuous education. Within CROMBES, there are three professional sections: Clinical Engineering, Medical Physics, and Biomechanics Section, respectively. They are especially active within the clinical environment. CROMBES is also a member of international organizations: IFMBE, IOMP, EAMBES, ESEM, and EFOMP. Since 2002, CROMBES has annually presented award for the best diploma paper in the interdisciplinary field of biomedical engineering.

Participants of MEDICON 2001

Hrvatsko društvo za medicinsku i biološku tehniku

Hrvatsko društvo za medicinsku i biološku tehniku objedinjuje više od 100 znanstvenika i stručnjaka koji djeluju u području biomedicinskog inženjerstva, a osnovna djelatnost im je u različitim granama znanosti. HDMBT je bio, a nadamo se da će i ostati glavna platforma za interdisciplinarnu suradnju u području BME. Članovi HDMBT-a sudjeluju u različitim međunarodnim i domaćim projektima i programima. Posebno ćemo istaknuti projekt *Neinvazivna mjerenja i postupci u biomedicini.* koji temeljen na suvremenim tehnologijama obećava ne samo daljnja istraživanja i razvoj nego i primjenu dostignuća u kliničkoj praksi te njihovo iskorištenje u gospodarstvu. Nastava u području biomedicinskog inženjerstva počela je u Hrvatskoj u ranim 70-tim godinama, u području biomedicinske elektronike na Sveučilištu u Zagrebu Fakultetu elektrotehnike i računarstva.

Danas se nastava iz biomedicinskog inženjerstva odvija u više sastavnica Sveučilišta u Zagrebu, a prisutna je i u ostalim sveučilišnim centrima u Hrvatskoj.

U okviru reforme visokog školstva u Hrvatskoj, planiramo osnovati studij biomedicinskog inženjerstva na *master* razini tako da se nastava odvija na više sastavnica sveučilišta, ali tim projektom namjeravamo također pojačati mobilnost nastavnika i studenata.

Ova nastojanja poduprta su i međunarodnim Tempus projektom. U proteklom desetljeću HDMBT je značajno unaprijedio međunarodnu suradnju na istraživačkim projektima sa stranim sveučilištima. Također organizirali smo više međunarodnih znanstvenih skupova poput 8th International IMEKO Conference on Measurement in Clinical Medicine, Dubrovnik 1998; 9th Mediterranean Conference on Medical and Biological Engineering and Computing-MEDICON 2001, Pula 2001; 1st International Summer School: Applications of ICT in Biomedicine, Dubrovnik 2002; i u suradnji s medicinskim fizičarima 2nd Austrian, Italian, Slovenian and Croatian Medical Physics Meeting-AISCMP, Opatija 2006. HDMBT će također aktivno sudjelovati u organizaciji 17th International Conference on Biomagnetism, u Dubrovniku, proljeće 2010.

HDMBT aktivnom radi na promociji kliničkog inženjerstva i u suradnji s IFMBE radni na uvođenju registriranja i ovlašćivanja kliničkih inženjera kao i na izgradnji programa njihovog trajnog obrazovanja. HDMBT također okuplja i medicinske fizičare u svojstvu njihove stručne udruge. HDMBT je član međunarodnih organizacija: IFMBE, IOMP, EAMBES, ESEM i EFOMP.

HDMBT od 2002. g. na svojoj godišnjoj skupštini dodjeljuje nagradu za najbolji diplomski rad u interdisciplinarnom području biomedicinskog inženjerstva.

The Cuban Bioengineering Society (SCB)

SOCBIO
Sociedad Cubana de Bioingeniería

The Cuban Bioengineering Society (SCB) was founded in 1978, by a group of engineers, mainly from the clinical areas of the Cuban Health System (CHS). At present, it has broadened its membership with graduates working in the CHS, universities, research and development institutions, manufacturers, quality control, and certifying institutions, as well as other national institutions. The SCB has accomplished a wide scope of activities, acknowledged locally as well as internationally. The landmark was the II Latin-American Congress of Biomedical Engineering, held in 2001, in Havana, with more than 600 participants, nearly half of them from other countries and the rest from Cuba itself. Since its establishment the SCB has beens affiliated to the IFMBE and was a co-founder of the Regional Biomedical Engineering Council for Latin America (CORAL).

The SCB has maintained a systematic international activity during the last 15 years. Its members have taken part uninterruptedly in the World Congresses of Medical Engineering and Medical Physics since 1994, as well as in the IFMBE meetings. SCB members have taken part in all the Latin-American congresses since their very beginning (Mazatlán 98) to date, while SCB officers have been present in all periodic CORAL meetings, maintaining an active work schedule in this organization, reinforced in the period 2004-2007, in which a member of SCB was CORAL's President. Efforts have been devoted several times to the creation of an EMBS national chapter that was completely impossible due to the anti-Cuban policy of the United States.

Quite an important number of SCB members are part of a health system built upon a principle of universal coverage of medical services. During the last years, the institutions supporting this health system have made outstanding investments in medical equipment intended for Primary Health Care as well as for improving the hospital system. The strengthening of the import of biomedical engineering in Primary Care is a distinctive characteristic in Cuba. Cuban health professionals have undertaken medical support in more than 80 countries that have asked for it. As a rule, biomedical engineers, members of the SCB, form part of the groups of medical and paramedical specialists in these countries. The rapid assembly of 30 field hospitals to assist Pakistani victims after the 2005 earth quakes in this country is a clear example of this support.

Officers of the SCB have systematically organized and taken part in national activities and have also supported different ones that have been held in different regions of the country. Thus, national SCB congresses held in alternate years, workshops between congresses, sponsoring and organizing congresses for third parties as research or health institutions, universities or other societies that have asked for collaboration are an outstanding sample of the SCB activity. Supporting the universities in Cuba, it played an important role in laying the base for the creation of Biomedical Engineering studies as an undergraduate career. As a result, the Biomedical Engineering career was created and initiated in three of the most important universities in the country in 2005.

During the next few years, it is foreseen that an increasing number of professionals will be engaged in the design of medical equipments and systems, supporting the continuously increasing development of health care in Cuba and its international importance. An important achievement will be organizing the V Latin American Congress of Biomedical Engineering, Havana 2011 (http://www.habana2011.sld.cu).

Sociedad Cubana de Bioingeriería

La Sociedad Cubana de Bioingeniería, fundada 1978, por un grupo de ingenieros de las áreas clínicas del sistema de salud cubano, ha extendido su membresía con profesionales de las instituciones del sistema de salud, universidades, centros de investigación y desarrollo, centros de producción, instituciones de registro, control de la calidad y otras instituciones nacionales. La organización ha logrado un alto nivel de actividad, reconocido nacional e internacionalmente, cuyo hito fue el II Congreso Lationamericano en que el participaron más de 300 delegados extranjeros e igual número de cubanos. Desde su fundación se afilió a la IFMBE.

La Sociedad ha tenido una actividad internacional constante en los últimos quince años. Se destaca la participación ininterrumpida de sus miembros en los congresos mundiales de Ingeniería Biomédica y Física Médica desde 1994 y asambleas de la federación, la participación en los congresos latinoamericanos de forma ininterrumpida desde sus inicios (Mazatlán´1998), y la asistencia a todas las reuniones periódicas de Consejo Regional de Ingeniería Biomédica para América Latina y la participación activa en el trabajo esta organización especialmente en el período 2004-2007 que fue presidido por un miembro de nuestra Sociedad.

Sus miembros forman parte de un sistema de salud basado en el principio de cobertura universal de sus servicios. En los últimos años el sistema ha hecho inversiones muy significativas en el equipamiento biomédico de la Atención Primaria de Salud y su sistema hospitalario. El fortalecimiento de la actividad de la ingeniería biomédica en la atención primaria es un rasgo distintivo de la actividad en el país. Los profesionales cubanos de la salud ofrecen también colaboración médica en más de 80 países que han solicitado estos servicios y junto a los médicos y enfermeras casi siempre están presentes ingenieros biomédicos cubanos que forman parte de esta asociación. El montaje de 30 hospitales de campaña para asistir a las víctimas del terremoto ocurrido en el año 2005 en Pakistán es un ejemplo destacado de ello.

La Junta de Gobierno ha trabajado sistemáticamente tanto en actividades nacionales como en el apoyo a las acciones que se realizan en diversas regiones del país. En este sentido se destaca la celebración de los congresos nacionales con carácter bienal, la realización de talleres nacionales entre ediciones de eventos, la participación como auspiciadores y organizadores de eventos realizados por universidades, institutos de investigación, instituciones de salud u otras sociedades en las que se nos ha convocado a participar. Junto a las universidades del país, trabajó en la fundamentación y solicitud a las autoridades nacionales de la inclusión de la Ingeniería Biomédica en el nivel de pregrado que dio como resultado su apertura en las tres universidades técnicas más importantes del país en el año 2005.

En los próximos años, se prevé que un número aún superior de profesionales trabajará vinculado al diseño de equipos y sistemas de aplicaciones médicas en todo el país dado el continuo desarrollo de la salud y su proyección internacional, lo que exigirá un esfuerzo particular de la Sociedad. Un hito especial en esa dirección será la organización del V Congreso Latinoamericano de Ingeniería Biomédica, Habana 2011.

Cyprus Association of Medical Physics and Biomedical Engineering (CAMPBE)

The Cyprus Association of Medical Physics and Biomedical Engineering, CAMPBE for short, was created in 1988.

It is an "umbrella" type organisation encompassing Medical Physicists, Biomedical and Hospital Engineers, Medical Doctors and other professionals whose work is related to Physics or Engineering in Medicine.

Since its creation CAMPBE has been promoting, developing and disseminating technology and science related to Medical Physics and Biomedical Engineering as well as enhancing Radiation Protection by promoting relevant knowledge, information and training for the benefit of the public.

Interdisciplinary collaboration and activities are fostered by organising every year a conference and by running training courses on radiation protection. Members are kept informed by the Association's Newsletters the "CAMPBE News".

CAMPBE honours its international commitments and two delegates have represented it at the World Congresses of Medical Physics and Biomedical Engineering that take place every 3 years since 1991.

Two major international conferences were successfully held in Cyprus, both were co-organised by CAMPBE and the Computing Department of the University of Cyprus.

Biomedical Research Foundation (BRF) is the host of the on-going research activities of the Cyprus Association of Medical Physics and Biomedical Engineering.

Research Projects (Completed)

1. 1992

Natural Radon measurement programme in Cypriot houses, which was finalised in January 1992. The results were published in international scientific journals and also were announced in the Cypriot press and radio.

2. 1994

Management of Medical Equipment – MED CAMPUS Project 401. This was a European Project with the objective to provide appropriate stimulus and inputs at national, as well as regional levels, in order to facilitate the necessary changes and promote the required awareness pertaining to the management, maintenance and repair of health care equipment. The activities under this project took place all through 1994.

3. 1994

Annual Dose Equivalent to the Cypriot Population from Natural Sources. The results of this research Project were presented at the World Congress of Medical Physics held in Rio de Janeiro, Brazil, in August 1994.

4. 1996-1998

Correlation between PSA level and evidence of Metastasis in Bone Scintigrams (1996 – 1998), as a Coordinated Research Project of the International Atomic Energy Agency. The aim of this project was to establish the physiological levels of PSA in the Cypriot Population and to investigate if there is a correlation between the PSA level in pathological patients and the indication of metastasis in their Bone Scintigrams.

5. 1998

Acceptance Testing and Calibration of the Medical Equipment of the Bank of Cyprus Oncology Centre. It was carried out in 1998. The results and experiences were presented in a Radiotherapy Oncology Meeting held in Cyprus in 1999.

Research Projects (Ongoing)

1. Quality Assurance of X-ray Diagnostic Equipment of the Cypriot Private Sector.

2. SENTINEL. This is a European Network Project for establishing Quality Assurance Procedures for Digital X-ray Equipment. BRF is one of its 23 collaborating partners.

Περίπου CAMPBE

Ο Σύνδεσμος Ιατρικής Φυσικής και Βίο-Ιατρικής Μηχανικής Κύπρου, CAMPBE δημιουργήθηκε το 1988.

Είναι μια οργάνωση που καλύπτει τους Φυσικούς Ιατρικής, τους Βίο-Ιατρικούς Μηχανικούς, Ιατρούς και άλλους επαγγελματίες των οποίων εργασία συσχετίζεται με τη φυσική ή την εφαρμοσμένη μηχανική στην ιατρική.

Η δημιουργία του CAMPBE έχει προωθήσει, αναπτύξει και διαδώσει στην Κύπρο, την τεχνολογία και την επιστήμη που αφορά την ιατρική φυσική και τη Βίο-Ιατρική μηχανική. Έχει επίσης συμβάλλει στην προστασία από τη ραδιενέργεια και την ακτινοβολία με την προώθηση σχετικής γνώσης, πληροφοριών και εκπαίδευσης προς όφελος του κοινού.

Διεπιστημονικές συνεργασίες και δραστηριότητες ενθαρρύνονται με την οργάνωση κάθε έτους ενός Συνεδρίου και με την διοργάνωση εκπαιδευτικών μαθημάτων για την προστασία από τη ραδιενέργεια και ακτινοβολία. Τα μέλη ενημερώνονται από την ιστοσελίδα του Συνδέσμου.

Το CAMPBE αντιπροσωπεύεται στα παγκόσμια συνέδρια της ιατρικής φυσικής και της Βίο-Ιατρικής εφαρμοσμένης μηχανικής που πραγματοποιούνται κάθε 3 έτη από 1991.

Δύο σημαντικά Διεθνή Συνέδρια έγιναν επιτυχώς στην Κύπρο, σε συνεργασία με το Πανεπιστήμιο της Κύπρου.

Το Ίδρυμα Βίο-Ιατρικών Ερευνών (ΙΒΕ, or Biomedical Research Foundation BRF), είναι θυγατρικός οργανισμός του Συνδέσμου και ασχολείται αποκλειστικά με την έρευνα.

Ερευνητικά Προγράμματα που έχουν ολοκληρωθεί

1. 1992

Πρόγραμμα μέτρησης ραδονίου σε οικίες της Ελεύθερης Κύπρου, το οποίο ολοκληρώθηκε τον Ιανουάριο του 1992. Τα αποτελέσματα δημοσιεύθηκαν στα διεθνή επιστημονικά περιοδικά και επίσης στον κυπριακό Τύπο και το ραδιόφωνο.

2. 1994

Διαχείριση του ιατρικού εξοπλισμού MED CAMPUS Project 401. Αυτό ήταν ένα ευρωπαϊκό πρόγραμμα με στόχο να παρασχεθεί το κατάλληλο ερέθισμα σε εθνικό, καθώς επίσης και περιφερειακό επίπεδο, προκειμένου να διευκολυνθούν οι απαραίτητες αλλαγές και να προωθηθεί η απαραίτητη συνειδητοποίηση σχετικά με τη διαχείριση, τη συντήρηση και την επισκευή του εξοπλισμού υγειονομικής περίθαλψης

3. 1994

Μέτρηση της Ισοδύναμης Ετήσιας Δόσης του κυπριακού πληθυσμού από τις φυσικές πηγές. Τα αποτελέσματα αυτού του ερευνητικού προγράμματος παρουσιάστηκαν στο παγκόσμιο συνέδριο της ιατρικής φυσικής που διοργανώθηκε στο Rio De Janeiro, Βραζιλία, τις 1994, Αυγούστου

4. 1996-1998

Correlation between PSA level and evidence of Metastasis in Bone Scintigrams (1996 – 1998), με τη συνεργασία του Διεθνούς Οργανισμού Ατομικής Ενέργειας. Ο στόχος αυτού του προγράμματος ήταν να καθιερωθούν τα φυσιολογικά επίπεδα της PSA στον κυπριακό πληθυσμό και να ερευνηθεί εάν υπάρχει συσχετισμός μεταξύ του επιπέδου PSA στους παθολογικούς ασθενείς και της ένδειξης για μετάσταση στα σπινθηρογραφήματα οστών .

5. 1998

Έλεγχος για παραλαβή και χρήση του ιατρικού εξοπλισμού του Ογκολογικού Κέντρου της Τράπεζας Κύπρου. Πραγματοποιήθηκε το 1998. Τα αποτελέσματα παρουσιάστηκαν σε ένα Συνέδριο της Ογκολογίας και Ακτινοθεραπείας που πραγματοποιήθηκε στην Κύπρο σε 1999,

Ερευνητικά προγράμματα (τρέχοντα)

1. Ποιοτικός έλεγχος διαγνωστικού ακτινολογικού εξοπλισμού στον Ιδιωτικό Τομέα της Ελεύθερης Κύπρου

2. SENTINEL. Ευρωπαϊκό πρόγραμμα την καθιέρωση των διαδικασιών εξασφάλισης ποιότητας ψηφιακών ακτινογραφικών συστημάτων. Το ΙΒΕ είναι ένας απ

Society for Biomedical Engineering and Medical Informatics Affiliated to the Czech Medical Association of Jan Evangelista Purkyne (SBMEMI CMA JEP)

The Society has been founded in 1978 with the aim to support scientific and research activities and to increase expert knowledge among its members in the field of biomedical engineering and medical informatics as well as to provide information on results of its activities and their implementation in practice among expert community and lay public alike.

Important meetings organised by the Society include the following events:

- *1st East-European Regional Conference on Biomedical Engineering,* 1987
- Regional Middle- and East-European IFMBE Conference *Biomedical Engineering*, 1991
- Regional IFMBE Conference of the Danube countries *BIOSIGNAL 92*, 1992
- International WHO seminar *Rehabilitation Engineering, Technology and Assessment*, 1995
- International Conference *Information Technology Applications in Biomedicine* ITAB´97, 1997
- International IFMBE Conference and 3rd European Medical and Biological Engineering Conference EMBEC'05, 2005
- Co-organising biennial international conferences *BIOSIGNAL* in Brno

The Society has currently three specialised sections – Biophysics, Medical Informatics, and Telemedicine & e-Health.

The Society also guarantees publication of the BME-oriented journal Lékař a technika (Physician and Technology). It is the oldest regularly appearing journal focused on new knowledge in the area of technological research, development and application in health care. Since 1970, it brings both popular science and peer-reviewed contributions from Czech as well as foreign authors, presenting up-to-date and newly emerged findings in the field of medical technology. The journal mission is to bring the information namely to medical and non-medical professionals, to students in health care and medicine and to other disciplines active in health care.

Closing session - awarding winners of Young Investigators Competition (one of the winners - Rodica Cristescu)

Individual members of the Executive Committee work as Society representatives in different bodies and agencies covering fields like legislation, professional qualification, accreditation of educational programmes and specialities, all this in co-operation with representatives of other societies affiliated with the Czech Medical Society J.E.P.

The Society regularly organises meetings focused on Telemedicine & e/Health and Medical Informatics use in practice. In the year 2008, the Society also co-organised meetings called *Intensive Care Units in 21st Century, Quality Safety and Legislation Meeting,* and a Meeting of the IEC/TC87 WG8, WG9, WG13 and WG14 dealing with technical normalisation in the field of ultrasonic systems.

The Society is a member of the following international organisations: IFMBE, IMIA, and EFMI.

Society of web page http://www.sbmili.cz.

Společnost biomedicínského inženýrství a lékařské informatiky České lékařské společnosti Jana Evangelisty Purkyně (SBMILI ČLS JEP)

Společnost byla založena v roce 1978 a hlavním cílem bylo zajistit podporu vědeckovýzkumné činnosti a zvyšování úrovně odborných znalostí svých členů v oblasti biomedicínského inženýrství a lékařské informatiky a pro vytváření podmínek pro seznamování odborné i laické veřejnosti s jejími výsledky a pro jejich uplatňování v praxi.

Mezi významné mezinárodní akce, které společnost pořádala, patří následující:

- *1. východoevropská regionální konference biomedicinského inženýrství*, 1987
- Regionální konference IFMBE zemí střední a východní Evropy *Biomedical Engineering*, 1991
- Regionální konference IFMBE podunajských států *BIOSIGNAL 92*, 1992
- Mezinárodní seminář WHO *Rehabilitation Engineering, Technology and Assessment*, 1995
- Mezinárodní konference *Information Technology Applications in Biomedicine* ITAB´97, 1997
- Mezinárodní IFMBE konference 3rd European Medical and Biological Engineering Conference EMBEC'05, 2005
- každé 2 roky je spolupořadatelem mezinárodní konference BIOSIGNAL v Brně

Společnost má v současné době tři sekce, kterými jsou sekce Biofyziky, sekce Lékařské informatiky a sekce Telemedicíny & e-Zdravotnictví.

Společnost je také odborných garantem vydávání odborného časopisu Lékař a technika. Je nejstarším odborným pravidelně vycházejícím časopisem zaměřeným na nové poznatky výzkumu, vývoje, výroby a použití techniky ve zdravotnictví. Od roku 1970 přináší populárně vědecké a recenzované příspěvky českých a zahraničních autorů, kteří prezentují aktuální a nově dosažené poznatky v oblasti lékařské techniky. Posláním časopisu je tyto poznatky prezentovat zejména lékařům a technickým nelékařským pracovníkům pracujícím v lékařských zařízeních, studentům zdravotnických oborů a dalším pracovníkům ve zdravotnictví.

Closing session

Jednotliví členové výboru společnosti působí jako zástupci v různých orgánech a podílí se na činnostech v oblastech jako: legislativa, kvalifikace, akreditace studijních programů a oborů, spolupráce s ostatními odbornými společnosti v rámci ČLS JEP atd.

Pravidelně pořádá společnost odborné akce zaměřené na Telemedicínu & e-Zdravotnictví a na Lékařskou informatiku v praxi. Společnost byla v roce 2008 též spolupořadatelem akcí s názvem *JEDNOTKY INTENZÍVNÍ PÉČE v 21. století, kvalita, bezpečnost, legislativa* a Meeting of IEC/TC87 WG8, WG9, WG13 and WG14 zaměřené na technickou normalizaci v oblasti ultrazvukových systémů.

Společnost je členem následujících mezinárodních organizací: IFMBE, IMIA a EFMI.

www. stránka společnosti.cz
www.sbmili.cz

The Danish Society for Biomedical Engineering (DMTS)

The Danish Society for Biomedical Engineering (DMTS) is an association of engineers, physicians, and technicians from research, education and health institutions who, together with colleagues from industry, join in a common interest to utilize and develop biomedical technology.

The mission for the Society is to promote the scientific and technological development and utilization of biomedical engineering (BME) in Denmark. This is done by promoting contact between BME-professionals, by disseminating knowledge between colleagues via meetings and seminars, by collaboration with fellow societies in the Nordic-Baltic region and worldwide, and by in general to work for a broader visibility for BME in the society.

The Danish Society for Biomedical Engineering was established in 1973 and counts to day 403 individual and 66 industry members. The Society publishes 6 time per year its own periodical, (*Medicinsk Teknologi & Informatik*, in collaboration with two other societies and a publishing company.

The Society arranges meetings on selected BME-topics all over Denmark and is hosting an annual 2-day meeting gathering approx. 300 BME-professionals. DMTS is affiliated to the International Federation for Medical and Biological Engineering (IFMBE) and the European Alliance for Medical and Biological Engineering & Science (EAMBES)

Compared to the size of the country (5.5 mill), BME is relativeLY big business in Denmark. Danish BME companies develop, produce, and market more than 40 billion Dkr (US$7 billion, 2007) worldwide. That is equivalent to about 31% of the turnover in the Danish health sector. More than 40,000 people are employed in the Danish BME industry in Denmark and abroad.

Denmark has several university programs leading to degrees in biomedical engineering. In 2000, Aalborg University (AAU) was the first university in Northern Europe to start a full five year MSc BME curriculum. AAU now offers the BSc, MSc and PhD in BME and a master in Health Informatics. BME research at AAU includes tissue engineering, rehabilitation technology, neuro technology, cardiac technology, and medical informatics.

Aarhus University offers the BSc, MSc and PhD in BME. BME research in Aarhus includes research in medical imaging and cardiovascular technology. The Technical University in Denmark in collaboration with University of Copenhagen offers the BSc, MSc and Phd in BME and BME-research includes research in medical ultrasound and image analysis.

Kim Dremstrup	Benedikte Kruuse Lindvig	Per Overgaard Rasmussen
Chairman	Vice-Chairman	Secretary

Medicoteknik i Danmark

Dansk Medicoteknisk Selskab (DMTS) er en sammenslutning af ingeniører, læger og teknikere fra forsknings- og uddannelsesinstitutioner, sundhedsvæsenet og industrien med en fælles interesse i at anvende og udvikle medicinsk teknologi.

Formålet med selskabet er at fremme den videnskabelige og tekniske udvikling på medicotekniske områder i Danmark, ved at etablere kontakt mellem medicoteknisk professionelle, formidle viden via møder, seminarer o.lign., samarbejde med øvrige nordiske og internationale selskaber for medicoteknik, samt generelt at udbrede kendskabet til medicoteknik.

DMTS blev etableret i 1973 og har i dag 403 personlige medlemmer og 66 firmamedlemmer. Selskabet udgiver sit eget danske medlemsblad (*Medicinsk Teknologi & Informatik*) i samarbejde med Dansk Selskab for Medicinsk Informatik, Dansk Selskab for Klinisk Telemedicin og Scanpublisher A/S. Bladet udgives 6 gange årligt.

Selskabet afholder møder om relevante medicotekniske emner rundt om i landet og afholder hvert år et stort samlende landsmøde over to dage med ca. 300 deltagere. Selskabet er affilieret medlem af the International Federation for Medical and Biological Engineering (IFMBE) og the European Alliance for Medical and Biological Engineering & Science (EAMBES).

I forhold til landets størrelse spiller medicoteknik en stor rolle i Danmark. Danske medicoteknik virksomheder udvikler, producerer og sælger for mere en 36 milliarder DKr i hele verden. Mere end 35.000 mennesker er ansat i den danske medicotekniske industri i Danmark og worldwide (2007).

I Danmark udbydes der universitets programmer på bachelor, kandidat og phd-niveau i medicoteknik på universiteterne i Aalborg, København og Aarhus. Der forskes bredt i medicoteknik på alle universiteterne i Danmark, men specielt fokuseres der på rehabilitering, neurofysiologi, kardiologi, medicinsk informatik samt billeddiagnostiske systemer og metoder.

For mere information om DMTS se venligst selskabets hjemmeside: www.dmts.dk

Kim Dremstrup	Benedikte Kruuse Lindvig	Per Overgaard Rasmussen
Formand	Næstformand	Sekretær

Selskabets sekretær: Per Overgaard Rasmussen × Vester Søgade 68 st. th. × DK-1601 Kbh. V

Telefon.: +45 33120418× E-mail: porasmus@post9.tele.dk
Web: http://www.dmts.dk

European Alliance for Medical and Biological Engineering and Science (EAMBES)

The mission of EAMBES is to improve the health, wealth, and well-being of the people by the application of medical and biological engineering and sciences.

EAMBES is an international association whose activities extend throughout Europe, as defined by the European Region of the World Health Organisation (WHO).

The objectives of EAMBES are to:

- Serve and promote Medical and Biological Engineering and Science (MBES) at European and national levels.
- Foster, co-ordinate, and provide added value to the activities of member organisations in MBES, and collaborate where appropriate with national and international organisations, particularly the International Federation for Medical and Biological Engineering.
- Serve and promote MBES education, training, and accreditation programmes.
- Serve and promote MBES research and development.
- Establish recommendations for the appropriate general responsibilities, organisational relationships, and roles of those engaged in the field of MBES.
- Establish and maintain liaison with national and European governments and agencies.
- Promote public awareness of MBES.
- Improve intersociety relations and co-operation in Europe within MBES and related fields.
- Encourage and assist in the formation of organisations for MBES where such organisations do not exist.

To achieve these goals, EAMBES is actively involved in the following activities:

1. Organise expert brainstorming meetings in the MBES area.
2. Prepare reports on the status and impact of MBES upon the society.
3. Apply for, participate in, and manage projects related to the promotion of MBES.
4. Collaborate with European authorities, agencies, and organisations for the promotion of MBES and for the allocation of funding and professional opportunities for the EAMBES members and the MBES community at-large.
5. Promote international collaboration and exchange of knowledge by:
   - The establishment of R & D networks.
   - Sponsoring international conferences and other meetings.
   - Supporting the publication of reports, proceedings, and journals.

6. Contribute to the coordination (harmony) and development of BME curricula;
7. Identify and disseminate examples of good practice in BME;
8. Act as a source of expert opinion on MBES education, training, accreditation, and application.

Maastricht, December 2, 2008
Dick W Slaaf
Secretary-General EAMBES
http://www.eambes.org

EAMBES Council 2003.
(L-R: Jos Vander Sloten, J. Barbenel, N. Saranummim, H. Hutter, J. Nagel)

The European Society for Engineering and Medicine (ESEM)

Educating Students and Professionals in Bridging the Disciplines

ESEM is the only European Society for Biomedical Engineering with individual members.

ESEM focuses on medical students, Biomedical Engineering students, PhD's, researchers, and industrial specialists, active or interested in Biomedical Engineering. ESEM is active in educating students to bridge the medical and engineering disciplines. ESEM is active in lifelong learning of professionals in bridging the disciplines within Biomedical Engineering.

The activities of ESEM include:

1. Educating professionals in bridging health care and engineering by:
   - Organising engineering workshops at medical congresses, focused on a specific medical field; engineers see the problems in this particular field and medical specialists see which technologies are available to solve their problems.
   - Realising textbooks for Biomedical Engineering courses, like the *Primer on Engineering for Medicine* that contains basic engineering knowledge for medical specialists, meant to prepare them for an MSc in biomedical engineering.
2. Educating professionals in bridging the disciplines within the broad area of Biomedical Engineering by:
   - Organising workshops, focused on a specific medical field like ventilation, showing all available technologies and discussing future research.
   - Organising congresses showing the latest developments in the broad area of Biomedical Engineering.
   - Presenting new technologies in *Technology and Health Care*, the official journal of ESEM. It is a peer-reviewed scientific journal, published by IOS Press and cited in several abstracting services.
3. Educating students in bridging health care and engineering by:
   - Organising summer schools for medical and engineering students, focused on teamwork. Medical students follow basic engineering courses, engineering students follow basic medical courses, and mixed teams follow lessons in teamwork and methodical design and practice teamwork by performing a design assignment.
   - Offering a consortium of high-quality universities with excellent Bachelor's and Master's programs that facilitates the exchange of students.

- Offering a consortium of high-tech industries that facilitates students in finding a traineeship, Master's project, or job.

4. Educating students in bridging the disciplines within the broad area of Biomedical Engineering by:
    - Organising workshops where MSc- and PhD-students present their work and get feedback from the ESEM board members.
    - Organising a European Masters in BME.

For more information please visit our website: www.esem.org.

The Estonian Society for Biomedical Engineering and Medical Physics

ESTONIAN
SOCIETY
FOR
BIOMEDICAL
ENGINEERING
AND
MEDICAL PHYSICS

The Estonian Society for Biomedical Engineering and Medical Physics (ESBMEMP) was established in 1994, as a joint society. In small countries, such as Estonia, it is probably unreasonable to differ between educational specialists and professionals in Medical Physics (MP) and Biomedical Engineering (BME), because the possibilities of employment within so narrow a field of qualification are quite limited. Today, the total number of members is 60. The three main professional disciplines: biomedical engineering, medical physics, and medicine, are all represented in the Estonian Society. ESBMEMP is affiliated to IFMBE (International Federation for Medical and Biological Engineering), IOMP (International Organization for Medical Physics), EFOMP (European Federation of Organizations for Medical Physics), and EAMBES (European Alliance for Medical and Biological Engineering & Science).

Historically, ESBMEMP has played a coordinating role in the creation of the first institutional structures to support BME education and study programs that were organized in two main public Universities – the Tallinn University of Technology and the University of Tartu. In 2001, as a result of bilateral agreement between the Estonian Ministry of Social Affairs and ESBMEMP, a Strategic Plan for Development of Medical Technology, until the year 2015, was developed. This plan is a part of the Estonian Healthcare Project 2015, and the main idea of this document is to enhance the quality of medical service using up-to-date medical technology and quality assurance. An important part of this paper is the number and educational level of technical staff in related areas not only in hospitals but also in different institutional bodies connected with the health care system.

ESBMEMP plays an important role in the creation of a BME profession in Estonia, in cooperation with the Estonian Association of Engineers, which comprises all the major engineering institutions in Estonia and represents Estonian engineers in FEANI (European Federation of National Engineering Associations). Today, ESBMEMP has a special commission of experts who coordinates oversight of engineering professions in all levels, development of a system for Continuing Professional Development (CPD), and evolves cooperation between Universities and Industry in BME area. In last few years, we have built up an authorized 3 level BME professional system in Estonia. First level is Biomedical Engineer (Eng) based upon 4 years university study plus 1 year working experience as engineer or 3 years university study plus 3 year working experience as engineer. Second level is Diploma Biomedical Engineer (Dipl.Eng) based upon 5 years university study plus 2 year working experience as engineer plus participation in CPD. The highest level is Chartered Biomedical Engineer (Chart.Eng) based upon accreditation as Dipl.Eng plus 2 year working experience as Diploma Engineer plus participation in CPD.

The aim of the ESBMEMP has been promotion of science and technology in its field of interest and coordination of collaboration between physicians and technical experts. As our main mission in the future, we see further work to unite medical and technological know-how for the benefit of human health and well-being.

Participans od NBC 1999, Tallin, Estonia

Prof. Kalju Meigas
National Secretary of ESBMEMP

Eesti Biomeditsiinitehnika ja Meditsiinifüüsika Ühing

ESTONIAN SOCIETY FOR BIOMEDICAL ENGINEERING AND MEDICAL PHYSICS

Eesti Biomeditsiinitehnika ja Meditsiinifüüsika Ühing (EBMÜ) moodustati 1994 aastal ja ta ühendab nii insenere kui füüsikuid. Väikestes riikides, nagu Eesti, ei ole otstarbekas eraldada biomeditsiinitehnika ja meditsiinifüüsika erialasid ega nende koolitust, kuna vastavate spetsialistide tööturg on nii kitsastes valdkondades piiratud. Tänasel päeval on ühingul 60 liiget. Liikmeteks on insenerid, füüsikud ja arstid. EBMÜ on liitunud nii ülemaailmsete ühingutega IFMBE (International Federation for Medical and Biological Engineering), IOMP (International Organization for Medical Physics) kui ka Euroopa ühingutega EFOMP (European Federation of Organizations for Medical Physics) ja EAMBES (European Alliance for Medical and Biological Engineering & Science).

Ajalooliselt on EBMÜ koordineerinud biomeditsiinitehnika õpetamiseks vajalike struktuuride moodustamist ja õppekavade arengut kahes Eesti suuremas ülikoolis – Tallinna Tehnikaülikoolis ja Tartu Ülikoolis. 2001 aastal sõlmiti kahepoolne kokkulepe Eesti Sotsiaalministeeriumi ja EBMÜ vahel, mille tulemusel töötati välja Eesti meditsiinitehnika arengukava aastani 2015. See kava moodustab osa Eesti Tervishoiu Arengukavast aastani 2015 ja paneb pearõhu arstiabi kvaliteedi tõusule kasutades kaasaegset meditsiinitehnoloogiat ja süsteemset kvaliteedijälgimist. Dokumendi oluliseks osaks on ka meditsiinitehnilise personali vajadus ja haridustase nii haiglates kui ka teistes tervishoiuga seotud asutustes.

EBMÜ on mänginud olulist rolli biomeditsiiniinseneri kutse arendamisel ja vastavate nõuete väljatöötamisel koostöös Eesti Inseneride Liiduga, kes esindab Eestit FEANI-s (European Federation of National Engineering Associations). EBMÜ-s on moodustatud ja töötab ekspertide komisjon, mis koordineerib biomeditsiiniinseneride kutsete andmist kõigil tasemetel, samuti nende elukestvat õpet ning arendab ülikoolide ning tööstuse koostööd vastavas valdkonnas. Viimastel aastatel on üles ehitatud biomeditsiiniinseneride kolmeastmeline kutsete süsteem. Madalaima tasemega kutse on biomeditsiiniinsener (Eng), mille saamise eelduseks on 4 aastat ülikoolis õppimist ja üks aasta tööd insenerina või 3 aastat ülikoolis õppimist ja 3 aastat tööd insenerina. Järgmise taseme kutse on diplomeeritud biomeditsiiniinsener (Dipl.Eng). Selle saamise eelduseks on 5 aastat ülikoolis õppimist ja 2 aastat töötamist insenerina ning lisaks aktiivne erialane enesetäiendamine. Kõrgeim tase on Eesti Volitatud Biomeditsiiniinsener (Chart,Eng), mille eelduseks on diplomeeritud biomeditsiiniinseneri kutse omamine, lisaks 2 aastat tööd ning samuti aktiivne erialane enesetäiendamine.

EBMÜ peaeesmärgiks on alati olnud biomeditsiinitehnika ja meditsiinifüüsika alase teaduse ja tehnika arengu koordineerimine ning sellele kaasaaitamine. Lisaks meditsiini ja tehnika ekspertide vahelise koostöö tugevdamine. Oma põhilise tuleviku visioonina näeme jätkuvat tööd, et ühendada meditsiiniline ja tehnoloogiline teadmus ning kogemus inimeste tervise ja heaolu parandamiseks.

Opening ceremony NBC 1999, Tallin, Estonia

Kalju Meigas
EBMÜ sekretär

The Finnish Society for Medical Physics and Medical Engineering (LFTY)

The Finnish Society for Medical Physics and Medical Engineering is an interdisciplinary, non-profit, and professional society for physicists, engineers, and physicians. The Society was established in 1968, and the number of members at 2008 was about 140. The members of the Society are from educational and research institutes as well as from health care companies and hospitals. The purpose of the Society is to arouse and maintain interest towards medical physics, medical engineering, and biophysics in Finland. The Society also aims to strengthen the development of research, education, and industry in these disciplines.

Starting from the year 2002, the activity of the Society has included the arrangement of the yearly Medical Physics and Medical Engineering Day. This day is traditionally arranged in January or February in different university towns in Finland. A significant event during the day is a poster competition where the best Master's theses, graduated in the previous year, from the field of medical physics and medical engineering are rewarded.

The Society has also been active in the development of research related to medical physics and medical engineering in Finland. For example, the society played a significant role in the establishment of the graduate school called International Graduate School for Biomedical Engineering and Medical Physics, which started its activity in 2007.

Mika Tarvainen,
Secretary of the Finnish Society for Medical Physics and Medical Engineering,
secretary@lfty.fi

Poster session on Finnish BME Day

Lääketieteellisen fysiikan ja tekniikan yhdistys (LFTY)

Lääketieteellisen fysiikan ja tekniikan yhdistys (LFTY) on fyysikoiden insinöörien ja lääkäreiden yhteinen poikkitieteellinen, aatteellinen ja ammatillinen yhdistys. Yhdistys on perustettu vuonna 1968 ja yhdistyksen jäsenmäärä vuonna 2008 oli noin 140. Yhdistyksen jäsenistö kattaa lääketieteellisen fysiikan ja tekniikan koko kentän oppi- ja tutkimuslaitoksista terveydenhuollon yrityksiin ja sairaaloihin. Yhdistyksen tarkoituksena on herättää ja ylläpitää kiinnostusta lääketieteelliseen fysiikkaan, tekniikkaan ja biofysiikkaan sekä edistää näiden aineryhmien tieteellisen tutkimuksen, koulutuksen ja teollisuuden kehitystä.

Vuodesta 2002 lähtien yhdistyksen toimintaan on kuulunut vuosittaisen lääketieteellisen fysiikan ja tekniikan päivän (LFT-päivän) järjestäminen. LFT-päivä on perinteisesti järjestetty tammi-helmikuun vaihteessa eri yliopistopaikkakunnilla. Kyseisen päivän keskeisenä osana on posterikilpailu, jossa parhaat lääketieteellisen fysiikan ja tekniikan alalta edellisenä vuonna valmistuneet diplomi- sekä pro gradu -työt palkitaan.

Yhdistys on ollut myös aktiivinen lääketieteellisen fysiikan ja tekniikan alan tutkimuksen kehittämisessä Suomessa. Yhdistyksellä oli muun muassa merkittävä rooli uuden tutkijakoulun (*International Graduate School for Biomedical Engineering and Medical Physics*) perustamisessa. Kyseinen tutkijakoulu aloitti toimintansa vuonna 2007.

Mika Tarvainen,
Lääketieteellisen fysiikan ja tekniikan yhdistyks,
secretary@lfty.fi

The German Society for Biomedical Engineering (DGBMT) within VDE

DGBMT DEUTSCHE GESELLSCHAFT FUR BIOMEDIZINISCHE TECHNIK IM VDE

German scientists in the field of biomedical engineering, bioengineering, clinical engineering, and medical ICT are organized in the DGBMT. The Society now has about 2,000 members including physicians, engineers, natural scientists, and companies. The Society serves as an inter- and multidisciplinary platform that brings together people from basic research, R&D, industry, and medical applications.

Nucleus of Knowledge

More than 20 technical committees and working groups cover the key technologies that are the driving forces of innovation and progress in medicine and healthcare. The committees work as rows of a matrix structure with medical focus topics as columns. Much of the work is done in a close co-operation with medical societies (e.g., the network for molecular imaging with experts from five of the leading German societies working on that particular field). A couple of years ago, the DGBMT founded the VDE Initiative MikroMedicine with the support of the German Government. The Initiative mainly focuses on telemonitoring for chronic diseases and on implants. People from ministries, from politics, and from health insurance companies meet here with people from R&D and industry.

Networking and Connecting

The technical committees are the backbone of the DGBMT. Their knowledge management is done mostly by committee meetings, in project groups and via web-based team portals. The technical committees organize various events in order to spread information and to support knowledge transfer. In 2008, the DGBMT held 18 events with 3,077 participants in total. Typically the Annual Meeting of the DGBMT is the biggest event with around 750 participants. The events deal with a wide range of themes from patient safety (APS & MEK 2008 Umbrella Congress on Patient Safety and Medical Technology) to neural engineering (IFESS 2008 Conference on Functional Electrical Stimulation). As one of the hosts, the DGBMT has placed considerable effort in organizing the World Congress on Medical Physics and Biomedical Engineering 2009 in Munich.

Products and Services

Members of the DGBMT get the Society's bimonthly scientific journal, *Biomedical Engineering,* published by Walter de Gruyter for free. Along with the journal, they get the magazine, *Health Technologies*. About 10 studies and position papers have been published during the last 5 years, most of which have been developed by the technical committees. During the same period of time, the book series *Health Academy* has been further developed. In 2008, volume 12, *Health Care 2025* and volume 13, *Model Guided Therapy,* were published.

Government Relations

Some of the DGBMT's studies and position papers had an impact for new funding programs of the German Ministry of Education and Research (BMBF). The DGBMT is also an advisor for the BMBF. In 2005, the DGBMT completed a study on the "Situation of Medical Technologies in Germany in an International Comparison". Along with a detailed analysis of key technologies, medical technologies, markets, and the industry sector, as well as the impact of the health care system, the study identified future funding priorities. In November 2008, the latest study for the BMBF had been published: *Study on the Identification of Innovation Barriers in Medical Technology.* Since 2002, the DGBMT has run projects on accompanying research for various thematic programs of the BMBF. Here the DGBMT does knowledge management, standardization, communication, and makes contact with the medical societies as networking partners in order to foster the acceptance of innovations.

Cooperation

The DGBMT is a member of the IFMBE, the Continua Health Alliance, the EAMBES European Alliance for Medical and Biological Engineering & Science, and the DVMT Umbrella Association Medical Technology, Natural Sciences, Informatics.

Among some co-operations with medical societies the DGBMT is a co-operating member of the AWMF Association of the Scientific Medical Societies in Germany (153 specialty societies).

Awards

The DGBMT maintains an awards program to recognize the achievement of young scientists. There are currently two awards: The Klee Award is given for an outstanding contribution to the field of biomedical engineering and the Olympus Award is given for an outstanding contribution to the field of optical technologies in medicine. Both awards are given yearly to a young scientist (not older than 35 years) for a publication; and come with prize money of € 5,000 for each award. Furthermore, a purse of € 2,000 in total is given to undergraduate students for the best presentations at the annual meetings of the DGBMT.

Deutsche Gesellschaft für Biomedizinische Technik (DGBMT) im VDE

Deutsche Wissenschaftler, die im Bereich Biomedizinische Technik, BioEngineering, Klinische Medizintechnik oder medizinische Informations- und Kommunikationstechnik arbeiten, sind in der DGBMT organisiert. Die Gesellschaft zählt über 2.000 Ärzte, Ingenieure, Naturwissenschaftler und Unternehmen zu ihren Mitgliedern. Sie bildet eine inter- und multidisziplinäre Plattform, die Experten aus der Grundlagenforschung, aus der angewandeten FuE, aus der Industrie und aus der medizinischen Anwendung zusammenbringt.

Wissensmanagement

Über 20 Fachausschüsse und Arbeitsgruppen bearbeiten alle wichtigen Schlüsseltechnologien, die die Innovationen und den Fortschritt in Medizin und Gesundheitswesen vorantreiben. Die Fachausschüsse arbeiten dabei wie die Zeilen in einer Matrixstruktur gemeinsam an medizinischen Fokusthemen, die man sich als Spalten der Matrix denken kann. Viele dieser Arbeiten geschehen in einer engen Kooperation mit medizinischen Fachgesellschaften. So arbeiten im Netzwerk "Molekulare Bildgebung" beispielsweise Experten aus fünf der führenden deutschen Gesellschaften zusammen. Vor einigen Jahren gründete die DGBMT mit Unterstützung des Bundesministeriums für Bildung und Forschung (BMBF) die VDE Initiative MikroMedizin. Die Initiative beschäftigt sich hauptsächlich mit Telemonitoring für chronische Erkrankungen und mit Implantaten. Sie bringt Experten aus Ministerien, aus der Politik und von Kostenträgern mit Experten aus Forschung, Entwicklung und der Industrie zusammen.

Networking und Kommunikation

Die Fachausschüsse bilden das Rückgrat der DGBMT. Sie betreiben Wissensmanagement durch Treffen, in Projektgruppen und über webbasierte Zusammenarbeit in Teamportalen. Um Informationen zu verbreiten und den Wissenstransfer aktiv zu unterstützen, organisieren die Fachausschüsse jährlich zahlreiche Veranstaltungen. Im Jahr 2008 hat die DGBMT beispielsweise 18 Veranstaltungen mit zusammen 3.077 Teilnehmern durchgeführt. Die Jahrestagung der Gesellschaft ist dabei mit ca. 750 Teilnehmern zumeist die größte und wichtigste Veranstaltungen eines Jahres. Die Spannbreite der Themen in den Veranstaltungen reichte dabei von Patientensicherheit (APS & MEK 2008 Dachkongress Patientensicherheit und Medizintechnik) bis zu Neuroprothetik (IFESS 2008 Conference on Functional Electrical Stimulation). Als einer der Gastgeber bereitet die DGBMT zudem zur Zeit intensiv den World Congress on Medical Physics and Biomedical Engineering 2009 in München vor.

Produkte und Dienstleistungen
Die DGBMT-Mitglieder erhalten alle zwei Monate die wissenschaftliche Zeitschrift "Biomedizinische Technik", die vom Verlag Walter de Gruyter publiziert wird. Zusammen mit der Zeitschrift bekommen Sie das Magazin "Health Technologies". In den vergangenen fünf Jahren hat die DGBMT mehr als 10 Studien und Positionspapiere veröffentlicht. Diese wurden zumeist von den Fachausschüssen erarbeitet. Im gleichen Zeitraum wurde die "Health Academy" als wissenschaftliche Buchreihe für die Hochtechnologiemedizin fortgeführt. Im Jahr 2008 erschienen bereits die Bände 12 "Gesundheitswesen 2025" und 13 "Modellbasierte Therapie".

Kontakte zu Ministerien
Einige der Studien und Positionspapiere, die von der DGBMT veröffentlicht wurden, haben das BMBF zu Förderprogrammen angeregt. Die DGBMT berät das BMBF auf verschiedenen Ebenen. So wurde im Jahr 2005 die "Studie zur Situation der Medizintechnik in Deutschland im internationalen Vergleich" abgeschlossen. Zusammen mit einer Analyse der Schlüsseltechnologien, der Biomedizinischen Technik, der Branche und ihrer Märkte sowie dem Einfluss des Gesundheitswesens auf diese Faktoren gab die Studie dem BMBF Handlungsempfehlungen für seine Forschungsförderung. Im November 2008 wurde die jüngste Studie für das BMBF abgeschlossen; die "Studie zur Identifizierung von Innovationshürden in der Medizintechnik". Seit dem Jahr 2002 führt die DGBMT Begleitforschung in verschiedenen Förderschwerpunkten des BMBF durch. Die Aufgaben dieser Projekte liegen im Wesentlichen bei Wissensmanagement, Kommunikation, Standardisierung und der aktiven Einbindung der medizinischen Fachgesellschaften, um deren Akzeptanz für die entwickelten Innovationen zu fördern.

Kooperationen
Die DGBMT ist Mitglied der IFMBE, der Continua Health Alliance, der EAMBES European Alliance for Medical and Biological Engineering & Science und des DVMT Dachverband Medizinische Technik, Naturwissenschaften, Informatik. Neben engen Kooperationen mit einer Reihe von medizinischen Fachgesellschaften ist die DGBMT kooperierendes Mitglied der Arbeitsgemeinschaft der Wissenschaftlichen Medizinischen Fachgesellschaften (AWMF), in der 153 deutsche Gesellschaften zusammengeschlossen sind.

Preise
Die DGBMT unterstützt hervorragende junge Wissenschaftler mit Preisen. Der Preis der DGBMT aus der Stiftung der Familie Klee ist für das gesamte Themenfeld der Biomedizinischen Technik bestimmt. Der Olympus-Preis der DGBMT für optische Verfahren in der Medizin wendet sich an einen speziellen Kreis. Beide Preise werden jährlich an junge Wissenschaftler bis 35 Jahre für herausragende Publikationen verliehen; sie sind mit jeweils € 5.000 dotiert. Die besten studentischen Präsentationen auf der Jahrestagung der DGBMT werden mit Preisgeldern von insgesamt € 2.000 ausgezeichnet.

The Hellenic Society of Biomedical Technology (ELEBIT)

The Hellenic Society of Biomedical Technology (ELEBIT) was created in 1976, and its aim is to promote professional excellence in Biomedical Engineering. Over its more than 30 years of activity, the Society has organized tens of International and National Conferences, Symposia, Seminars, and other scientific events. The Mediterranean Conference, MEDICON 89, held in Patras, is probably the most well-known. It will be repeated next year as MEDICON 2010, in Chalkidiki. ELEBIT is also sponsoring the series of the European Symposia on Bio-Medical Engineering (ESBME) held every two years in Greece and is supporting the Postgraduate program on BME held since 1989, at the University of Patras, in collaboration with the National Technical University of Athens and 25 other European Universities.

Member of the IFMBE since the early 1980s, ELEBIT has actively participated in its Bodies, working Groups and Divisions and supported its initiatives. Taking into account the enormous progress, changes, and new challenges in the field of biomedical Engineering worldwide, the recently elected new administrative council of ELEBIT took the initiative to review the constitution and by-laws of the Society, in order to better respond to the new demands. The Society is planning to play a more important role on technical, scientific, educational, and training issues related to medical device technology and innovation. It will also try to influence political decisions concerning the importance of biomedical research programs at a national level, for the benefit of the biomedical engineering community and the society in general.

Prof. Nicolas Pallikarakis,
President of ELEBIT

Ελληνική Εταιρεία Βιοϊατρικής Τεχνολογίας (ΕΛΕΒΙΤ)

Η Ελληνική Εταιρεία Βιοϊατρικής Τεχνολογίας (ΕΛΕΒΙΤ) ιδρύθηκε το 1976 και στόχος της είναι η προώθηση της επαγγελματικής αριστείας στον τομέα της Βιοϊατρικής Τεχνολογίας. Στα περισσότερα από τριάντα χρόνια δράσης της, η εταιρεία έχει οργανώσει δεκάδες Διεθνή και Εθνικά Συνέδρια, Συμπόσια, Σεμινάρια και άλλες επιστημονικές εκδηλώσεις. Το Μεσογειακό Συνέδριο MEDICON 89, που πραγματοποιήθηκε στην Πάτρα, είναι πιθανότατα το πιο γνωστό. Θα επαναληφθεί του χρόνου ως MEDICON 2010 στη Χαλκιδική. Η ΕΛΕΒΙΤ επιχορηγεί επίσης τα Ευρωπαϊκά Συμπόσια Βιοϊατρικής Τεχνολογίας (ESBME), που οργανώνονται ανά διετία στην Ελλάδα, ενώ στηρίζει και το Μεταπτυχιακό Πρόγραμμα Σπουδών στη Βιοϊατρική Τεχνολογία, το οποίο λειτουργεί στο Πανεπιστήμιο της Πάτρας από το 1989, σε συνεργασία με το Εθνικό Μετσόβειο Πολυτεχνείο και 25 ακόμα Ευρωπαϊκά Πανεπιστήμια. Μέλος της IFMBE από τις αρχές της δεκαετίας του '80, η ΕΛΕΒΙΤ έχει συμμετάσχει ενεργά στα Όργανα, τις Ομάδες Εργασίας και τα Τμήματά της, και έχει στηρίξει τις πρωτοβουλίες της. Λαμβάνοντας υπ' όψιν την τεράστια πρόοδο, τις αλλαγές και τις νέες προκλήσεις στον τομέα της Βιοϊατρικής Τεχνολογίας παγκοσμίως, το πρόσφατα εκλεγμένο νέο διοικητικό συμβούλιο της ΕΛΕΒΙΤ πήρε την πρωτοβουλία να αναθεωρήσει το καταστατικό και τους κανονισμούς λειτουργίας της εταιρείας, προκειμένου να ανταποκριθεί καλύτερα στις νέες απαιτήσεις. Η εταιρεία σκοπεύει να διαδραματίσει στο μέλλον έναν ακόμα σημαντικότερο ρόλο σε τεχνικά, επιστημονικά, εκπαιδευτικά, καθώς και θέματα που σχετίζονται με την καινοτομία των ιατροτεχνολογικών προϊόντων. Θα προσπαθήσει επίσης να επηρεάσει πολιτικές αποφάσεις που αφορούν τη σημασία των ερευνητικών προγραμμάτων σε εθνικό επίπεδο, προς όφελος της βιοϊατρικής κοινότητας αλλά και της κοινωνίας εν γένει.

Καθ. Νίκος Παλληκαράκης,
Πρόεδρος ΕΛΕΒΙΤ

The Hong Kong Institution of Engineers – Biomedical Division

The Biomedical Division of the Hong Kong Institution of Engineers has been taking bold but steady steps to promote the professionalism and enhance the capability of local engineers in biomedical engineering development. With numerous activities being completed in the Session 2008/09, the Division has organised various activities covering a very broad topics contributing from our strategic composition of the Committee, consisting of active biomedical engineers from the medical device industry, academia, regulatory body, government and semi-government organizations, raising the profile of the Division.

With the on-going government funded project under Professional Services Development Assistance Scheme (PSDAS) titled "To promote the professionalism and enhance the competence of local engineers in facilitating the design and manufacturing of high value-added medical and healthcare devices", a two-day conference cum 1-day technical visit "BME2008 Hong Kong Biomedical Engineering International Conference" was organised from 23 to 25 October 2008, with participant headcount over 400. BME2008 served as an exchange forum for local and international biomedical engineers, scientists, researchers, practitioners, regulators and industrialists, facilitating the newest information interflow and promoting learning from overseas successes and thus stimulating industrial enhancement. The Conference has been brought to a successful close, with acclamation credited to the biomedical engineering academic and research achievements, both inspiring and rewarding, as well as celebration for our biomedical engineering professionalism.

In the past Session, the Division has been closely collaborating with our affiliations in organising various CPD seminars and trainings with their intelligence in different aspects. The Division has been working closely with various Divisions within HKIE, including the Electronics Division, the Information Technology Division, the Manufacturing and Industrial Engineering Division, the Young Members Committee and the Associate Members Committee. We have also actively co-organised various activities with IEEE-EMBS Hong Kong Chapter, Hong Kong Medical and Healthcare Device Industries Association and the Biomedical Engineering Alumni Association of HKPolyU to promote the exchange of information in biomedical field.

Followed by the successful formation of the Biomedical Engineering Discipline in Hong Kong in 2004, there has been three Biomedical Graduate Scheme "A" Training Programme being hosted by the Electrical and Mechanical Services Department (EMSD) of the HKSAR Government, Hong Kong Productivity Council (HKPC) and Hospital Authority (HA), to provide formal training to biomedical engineering graduates in Hong Kong. There has been a continued growth on the number of members in the Biomedical Discipline, as well as the biomedical engineering arena and medical device industry in Hong Kong. Such growth has been intensified by the aging population and the heightened demand for quality of healthcare services. Indeed, biomedical engineering has gradually migrated from a macroscopic and systemic perspective to a microscopic and cellular one. One example was the recent research and development of novel diagnostic technologies due to the pandemic threats of SARS and avian flu, which aims at overcoming the challenging demand for high sensitivity, specificity and minimal turn-around time, and thus more efficient infectious disease control.

The expeditious advancement of medical technologies has also brought with it a number of challenges, one of them was the concern over the safety and effectiveness of medical devices. The proposed regulation of medical devices and the associated regulatory impact assessment exercise conducted by the Government has raised public awareness and encouraged active involvement of professionals. The Biomedical Discipline would continue to serve as a platform for the exchange of views and ideas among its members and to channel their expert comments to the Government. Furthermore, with the coming statutory regulation of medical devices, the Discipline anticipates that there will be an increasing demand for qualified biomedical engineering professionals in both the public and the private sectors.

In the coming Session, the Biomedical Division would continue enhancing the connections with the industry and the medical practitioners, and fostering the training of high quality engineers, to boost the development of the biomedical engineering profession in Hong Kong.

Group Photo in Annual General Meeting of the Biomedical Division of the Hong Kong Institution of Engineers on 17 April 2009

The International Council on Medical and Care Compunetics (ICMCC)

The ICMCC Foundation (www.icmcc.org) focuses on the social, societal, and ethical implications of computing and networking (compunetics) in medicine and care.

The main goal of the foundation is to be the central meeting place for exchanging information on all aspects related to medical and care compunetics and for all those concerned, making information on medicine and care available to patients using compunetics as well as distributing information on the use of compunetics in medicine and care to patients and professionals.

It has done so in its 4 years of existence by organizing the annual ICMCC event, the proceedings of these conferences are published in the IOS Press - *Studies in Health Technology and Informatics* series (http://conferences.icmcc.org/).

ICMCC created the only internet knowledge center on patient record access (http://recordaccess.icmcc.org/); the ICMCC News Page has become one of the leading sources for medical and care ICT related news and scientific articles (http://articles.icmcc.org/wpc/).

In 2009, ICMCC will publish a book on Digital Homecare and officially publicize its proposed guidelines on patient record access to the WHO.

ICMCC Board and Speakers, 2005

Scientific Society for Measurement, Automation and Informatics (MATE); Biomedical Engineering Section

The Scientific Society for Measurement, Automation and Informatics was founded in 1952, with the aim to ensure opportunity for experts in the field of measurement, automation and informatics to share their knowledge.

The Biomedical Engineering Section specifies itself to be a knowledge center – both on scientific and educational level – for medical technology and medical devices. Our primary aim is to help and motivate knowledge transfer between experts and researchers in healthcare, industry and universities. To fulfill these aims our Section organizes annual meetings and is the co-organizer of the "BUDAMED" biomedical engineering conferences, which take place every third year (BUDAMED'08 was the 5th such event). We put a great emphasis on involving young members in our society as well as on propagating biomedical engineering, therefore, a Young Investigators Competition is organized annually for graduates.

Our Society has initialized cooperation with other scientific societies. We are convinced that small specific societies cannot survive.

IEEE Engineering in Medicine and Biology Society

IEEE Engineering in Medicine and Biology Society (EMBS) is the world's largest international society of Biomedical Engineers. Its 8,300 members reside in some 70 countries around the world, forming a true global connection, bringing together fascinating people, practices, information, ideas, opinion, and fellowship from one of science's fasting growing fields. The mission of EMBS is to advance the application of engineering sciences and technology to medicine and biology, promote the profession, provide global leadership for the benefit of its members and humanity by disseminating knowledge, set standards, foster professional development, and recognize excellence. Fifty-two percent of EMBS members are in academia, forty-six percent in industry, and two percent in government.

EMBS holds an Annual International Conference that focuses on eleven themes: Biomedical Signal Processing; Biomedical Imaging and Image Processing; Bioinstrumentation, Biosensors, and Bio-micro/nano Technologies; Bioinformatics, Computational Biology & Systems Biology; Cardiovascular & Respiratory Systems Engineering; Neural & Rehabilitation Engineering; Cellular and Tissue Engineering and Biomaterials; Bio-Robotics & Biomechanics; Therapeutic and Diagnostics Systems and Medical Device Design; Healthcare Information Systems and Telemedicine; Technology Commercialization, Industry, Education and Society. In addition to the Annual Conference, Special Topics Conferences are also held that include Neural Engineering and Biomedical Imaging. The conference features a student paper competition with finalists from all parts of the world. The annual conference is routinely held in different regions of the world with recent conferences held in Shanghai (2005), New York (2006), Lyon France (2007), and Vancouver Canada (2008), and will be held in Minneapolis (2009), and Buenos Aires, Argentina (2010).

EMBS Publications include *EMB Magazine*, and a number of IEEE Transactions: *Biomedical Engineering*, *Information Technology in Biomedicine*, *Neural Systems and Rehabilitation Engineering*, *Medical Imaging*, *Nanobioscience*, *Computational Biology and Bioinformatics*, *and Biomedical Circuits and Systems*.

EMBS has over 100 local chapters in 41 countries, including North America (69), Europe (19), Asia-Pacific (16), Latin America (9), and Middle East/Africa (2). These chapters bring direct service to our members, many of whom are unable to attend our conferences.

EMBS belongs to several umbrella societies that have societies as members including International Federation of Medical and Biological Engineering (IFMBE), European Alliance for the Medical and Biological Engineering and Science (EAMBES), Latin American Regional Council on Biomedical Engineering (CORAL), and American Institute for Medical and Biological Engineering (AIMBE). EMBS co-sponsors its annual meeting with local societies when possible.

Presentation of Awards to IFMBE Young Investigator Competition winners, at the Annual IEEE EMBS Conference, Shangai 2005

The Institute of Physics and Engineering in Medicine (IPEM)

(United Kingdom of Great Britain and Northern Ireland)

The IPEM was formed in 1995, by the merger of the Biological Engineering Society (bioengineering) and the Institute of Physical Sciences in Medicine (medical physics), which means that IPEM is Affiliated to IFMBE and IOMP (and to EAMBES and EFOMP).

The Biological Engineering Society (BES) was founded in the UK in 1960. IPSM, originally known as the Hospital Physicists' Association, had been founded some years earlier, in 1943, and IPEM has just celebrated its 65th birthday. The impetus for the formation of BES was the founding of IFMBE at an international conference held in Paris a year earlier, in 1959. BES membership included clinical engineers working in hospitals, medical and surgical colleagues working with applications of engineering in their own fields, and other engineers working in research and development in areas related to bioengineering.

A BES member, Dr Alfred Nightingale, Lecturer in Physics at St Thomas's Hospital Medical School and Chief Physicist at the hospital, was the first Editor of IFMBE's official journal, *Medical Electronics and Biological Engineering* (now *Medical and Biological Engineering and Computing*), which was first published in 1963. Dr Nightingale died suddenly in February 1963, at the age of 40 after editing only the first issue of the journal, but he is remembered by the annual "Nightingale Prize" for the article in the journal judged "best", originally awarded jointly by IFMBE and BES, and now sponsored by IPEM.

Alfred Nightingale was typical of many members of BES, who had their roots in physics but who found applications for their physics research in medicine through engineering.

In 1979, BES launched its own journal, the *Journal of Biological Engineering*, which was renamed in 1994, as *Medical Engineering and Physics*. Jack (W. J.) Perkins was its first Editor and is remembered through the "Perkins Prize".

The number of BES members with physics roots was always a driver towards joint activities with IPSM. A number of joint Annual Scientific Meetings were held and for a number of years a joint membership directory was published. The overlapping membership of the two organisations eventually led to the 1995 merger, with an initial significant fall in income from member subscriptions, because a number of dual members now paid only one subscription and, inevitably, some members were not comfortable in a bigger, broader organisation and did not renew.

However, IPEM has continued to grow. In 1997 the Association of Medical Technologists merged with IPEM, bringing both clinical physics technologists and clinical engineering technologists into IPEM membership. Subsequently there has been organic growth across all sections of membership, but especially amongst technologists, following IPEM's support for improved arrangements for their education and training and their professional recognition.

IPEM has been granted licences by the Engineering Council UK to award the designations Chartered Engineer (CEng), Incorporated Engineer (IEng) and Engineering Technician (EngTech), and to accredit educational courses and training programmes that prepare candidates for engineer registration.

IPEM is conscious that its core area for membership, across all disciplines, is in healthcare service provision and development, and that there are many working in healthcare industries, and in academe, in teaching, research, and development, who would add to the richness of provision for professional bioengineering in the UK, if they can be persuaded to join IPEM. IPEM is, therefore, developing links with industry through a new Industrial Liaison Group and has also joined the UK Engineering Professors' Council as a Corporate Member to improve contacts with academic bioengineers in university departments.

Internationally, IPEM now provides secretariat support to IFMBE, and, more recently, to EAMBES from its office in York and plans to strengthen further its European and world-wide links in future years. IPEM has always had a small membership outside the UK but, conscious of its increasing reputation as a learned society, especially through its official journals in medical engineering, medical physics and physiological measurement, but also of the need to nurture and support the development of national member organisations wherever they can be viable, IPEM has introduced a low-cost membership category of Overseas Affiliateship, which gives access to IPEM journals and other member benefits, and which is open only to those who are members of their own relevant national member organisations, and who remain so.

The Biomedical/Clinical Engineering Association of Ireland (BEAI)

The Biomedical/Clinical Engineering Association of Ireland, (BEAI) (www.beai.org), was established in 1992, by a group of hospital-based Clinical Engineering professionals who recognised a need for a scientific association that would meet the many different needs of a very vibrant profession. Membership of this voluntary association is open to individuals employed in both the public and private sectors, whose principal function is to provide a Biomedical/Clinical Engineering service. Corporate membership is also available for companies that provide a medical devices service. There are currently 50 Public Acute Care Hospitals and a growing number of Private Care Hospitals in Ireland. The vast majority of all these hospitals have their own in-house biomedical/clinical engineering services. These engineers provide clinical, technical, and managerial expertise in the identification, acquisition, and support of safe, effective, and economical technology as part of a healthcare professional team. They can have responsibility for patient safety, electro medical device technology from pre-tender to decommissioning, quality assurance, and the effective and efficient utilization of electro medical device technologies.

The aims of the BEAI are to

- Encourage and promote the professional development of Biomedical/Clinical Engineering personnel employed in the Health Care services and support infrastructure.

- Advance the science, technology, ethics, and art of Biomedical/Clinical Engineering through association, education, training, publications, and other materials.

- Facilitate co-operation and understanding among Biomedical/Clinical Engineering personnel and other health care professionals, hospitals, academia, vendors, and other organizations with an interest in Biomedical/Clinical Engineering.

These aims are achieved through the use of their website www.beai.org, through their publication of the *BEAI Spectrum Journal*, which is currently produced three times a year. The aims are also achieved by information sharing and networking at their Annual Scientific Meeting, attendance at international Meetings, and smaller educational meetings.

The Annual Scientific Meetings (usually held in late September/early October) organised by the BEAI, provide an opportunity for engineers both in the public and private sectors to come together to network, share knowledge and experiences, thereby allowing all to positively develop. This continuous professional development helps to ensure that resources such as people, time, money, and equipment are efficiently and effectively utilized for the continual improvement and enhancement of the health care system.

Many members of the BEAI are also involved with other national bodies and organisations such as:

The Irish Medicine Board (IMB) (www.imb.ie), who is the Competent Authority under the Medical Devices Directive for Ireland,

The Biomedical Engineering Division of Engineers Ireland (www.engineersireland.ie).

The Irish Medical Surgical Trade Association (IMSTA), www.imsta.ie

The Irish Medical Devices Association (IMDA) www.imda.ie

The Technical Committee 10 (TC10) of the Electro Technical Council Ireland (ETCI). http://www.etci.ie/technical/tc10.html

This involvement and networking with these other associations ensures that there is a continuous dissemination of knowledge and exchange of professional ideas through out the country. Members of the BEAI continuously work with and network with industry and academia to enhance their knowledge for the advancement of patient care.

Internationally, the BEAI is the Irish representative body for the International Federation of Medical and Biological Engineering (IFMBE) (www.ifmbe.org) Organisation. The Secretary of the BEAI attends the secretarial meetings of the IFMBE representing the interests of the Association at these meetings. The IFMBE recently endorsed the 2008 Annual Scientific Meeting of the BEAI.

The Israel Society for Medical and Biological Engineering (ISMBE)

A short History of the Society's Last ~20 Years

The Israel Society for Medical and Biological Engineering (ISMBE) was established in 1958, and has been since then the largest and most active society in the country, representing the Biomedical Engineers as well as the Clinical Engineers. While initially most members were engineers educated in other fields, who were attracted to the field, the initiation in 1968 of Graduate level studies in Biomedical Engineering towards M.Sc. and Ph.D. degrees (at the Technion), have transformed ISMBE to a higher professional level. Formal undergraduate programs (4 years studies towards a B.Sc. degree in Biomedical Engineering) have started in 1999, at 3 research universities and one college, and currently about 200 students graduate each year. These BME have transformed and rejuvenated the constituency of the Society. Recently, several more colleges have also started offering undergraduate BME programs.

The ISMBE has a name list of over 1000 members, but paying membership is fluctuating around 200-300, depending on the activity from year to year (with people tending to pay their dues only when attending the annual conference – and not on a regular yearly basis). The majority of the members are working in R&D in industry or within the universities, with only a minority of about 20% working in service and maintenance, within hospitals or with vendors who provide such services to the hospitals. While there have been some medium sized manufacturers of medical devices in operation for over 30-35 years, the last 14 years have seen a significant increase of interest and involvement in Biomedical-Medical devices/equipment, in research, development, and manufacturing. There are currently around 500 start-ups in this area, at different levels of operation.

The annual ISMBE conference is usually a one day meeting, with several sessions which include lecturers from academia, industry, and from hospitals. Oral presentations are given as well as posters. For many years, this annual meeting is run as part of a national conference that included many of the medical societies. The recent increase of interest in the Biomedical Engineering discipline and in the ISMBE annual conference has justified the organization of regional activities; thus, 5-6 half-day workshops are organized in different locations around the country, usually covering one specific topic, with representation from academia, industries, and VC's. The ISMBE is also organizing an annual Job Fair for the undergraduate students, in which industry representatives lecture about their products/companies, as well as interview students for future employment.

The ISMBE has been active in international organizations since it was established. The ISMBE organized and hosted the International Conference of the IFMBE (in association with the International Organization for Medical Physics) in Jerusalem, Israel, in 1979. The Society also organized and hosted the IFMBE Mediterranean countries regional conference MECOMBE'92 in Jerusalem. Many other Medical and Engineering Conferences were held in Israel.

In recent years, the ISMBE has intensified its activities as promoter of BME activities within the various national funding institutes – in an effort to establish BME as a stand-alone discipline. These efforts must continue, since currently BME is still regarded as "some activity between Biotechnology, Medicine and Engineering...". ISMBE is attempting to change this attitude. There are multiple bi-lateral research funding agreements between Israel and various countries (e.g., the USA, European, and other nations), and the ISMBE is attempting to increase the level of BME activities within these frameworks.

Dan Adam
Technion – Israel Institute of Technology, Haifa, ISRAEL

היסטוריה קצרה של האיגוד

דן אדם
הטכניון – מכון הטכנולוגי לישראל, חיפה, ישראל

האיגוד הישראלי להנדסה רפואית וביולוגית נוסד ב-1958 ומאז הוא האיגוד הגדול והפעיל ביותר המייצג את המהנדסים הביו-רפואיים וכן את המהנדסים הקליניים בישראל. בשנים הראשונות, רוב חברי האיגוד היו מהנדסים שלמדו מקצועות אחרים (מהנדסי אלקטרוניקה, מהנדסי מכונות, מהנדסי תעשיה וניהול) אשר התענינו בשטח, ולמדו תוך כדי עבודה. פתיחת תוכניות לימודי מוסמכים בהנדסה ביו-רפואית בשנת 1968 (תחילה בטכניון), לקראת התארים מגיסטר (.M.Sc) ודוקטור (.Ph.D), הוסיפה משמעותית לרמה המקצועית של פעילות האיגוד. לימודי תואר ראשון (תוכנית 4 שנתית לקראת התואר .B.Sc בהנדסה ביו-רפואית) החלו בשנת 1999 בשלוש אוניברסיטאות מחקר ובמכללה אחת, כאשר כיום מסיימים מדי שנה כ- 200 בוגרים. מהנדסים אלו שינו בצורה משמעותית והצעירו את אוכלוסיית חברי האיגוד. לאחרונה, מכללות נוספות פתחו תוכניות לימוד תואר ראשון בהנדסה ביו-רפואית.

לאיגוד רשימת חברים של למעלה מ-1000 חברים, כאשר מספר החברים המשלמים את דמי-החבר השנתיים נע בסביבות 300-200, בהתאם לפעילות באותה שנה (כאשר רוב האנשים נוטים לשלם את דמי-החבר כאשר הם משתתפים בכינוס השנתי – ולא על בסיס שנתי קבוע). רוב חברי האיגוד עובדים במחקר ופיתוח בתעשייה ובאוניברסיטאות, כאשר מיעוט של כ- 20% עובד במתן שירותים ותחזוקה, עפי"ר בבתי חולים ואצל ספקי שירותים המספקים שירותים אלו לבתי החולים. בישראל קיימות חברות, בינוניות בגודלן, העוסקות בפיתוח וייצור של מיכשור רפואי זה למעלה מ- 35-30 שנה; אולם ב-15-14 השנים האחרונות חלה עליה משמעותית בפעילות מחקרית, פיתוח וייצור של מערכות, ציוד ומיכשור רפואיים וביו-רפואיים. קיימות כיום כ-500 חברות הזנק בשטח זה, ברמות שונות של פעילות.

הכינוס השנתי של האיגוד הוא בדרך כלל כינוס של יום אחד, הכולל הרצאות הניתנות ע"י אנשי תעשיה, אקדמיה ובתי חולים. עפי"ר מתקיימות הרצאות פרונטאליות והצגה של פוסטרים. זה מספר שנים שהכנס השנתי מתקיים כחלק מכינוס ארצי הכולל גם כנסים של אגודות רפואיות שונות, כך שאפשרית הפריה הדדית ומעבר לשמיעת הרצאות גם בשטחים רפואיים נוספים. ההתעניינות הגוברת בשנים האחרונות בשטח ההנדסה הביו-רפואית ובכינוס השנתי של האיגוד, הצדיקו אירגון וקיום של פעילויות אזוריות, 6-5 סדנאות של חצי יום, המוקדשות לנושא ספציפי, ובהן משתתפים נציגים מהאקדמיה, מהתעשייה ומחברות הון-סיכון. הארגון גם מארגן פעם בשנה "שוק תעסוקה" לסטודנטים הלומדים לימודי תואר ראשון, כאשר נציגים מהתעשייה מרצים על המוצרים /מכשירים המפותחים על ידם, וגם מראיינים סטודנטים לקראת אפשרות העסקתם מיידית או בעתיד.

האיגוד היה פעיל במסגרות ארגוניות בינלאומיות מאז היווסדו. האיגוד אירגן ואירח את הכינוס העולמי השנתי של ה-IFMBE (ביחד עם הארגון הבינלאומי לפיסיקה רפואית) בירושלים בשנת 1979. כמו כן האיגוד אירגן ואירח את הכנס האיזורי של מדינות אגן הים התיכון של ה-IFMBE - MECOMBE'92, אשר נערך בירושלים ב-1992. כנסים רבים נוספים בנושאים רפואיים-הנדסיים קוימו בחסות ובעזרת האיגוד.

בשנים האחרונות הגביר האיגוד את פעילותו לקידום הפעילויות בנושאי ההנדסה הביו-רפואית במסגרות המוסדות הלאומיים התומכים וממנים מחקר ופיתוח – במטרה לבסס את ההנדסה הביו-רפואית כדיסציפלינה עצמאית. מאמצים אלו חייבים להימשך, כיוון שגם כיום ההנדסה הביו-רפואית מוגדרת לפעמים כ- "פעילות שהיא נופלת בין ביוטכנולוגיה, רפואה והנדסה...". האיגוד מנסה לשנות גישה זאת. ישנן מספר קרנות מו"פ דו-לאומיות (כגון קרן שת"פ עם ארה"ב, עם אירופה, ומדינות נוספות), והאיגוד מקדיש מאמצים להגברת הפעילות של ההנדסה הביו-רפואית במסגרות אלו.

Italian Association of Medical and Biological (AIIMB)

President : **Prof. Antonio Pedotti**
Secretary General: **Prof. Marcello Bracale**

General Activities:
As early as 1965, a group of medical engineers, headed by Prof. G. Francini, met to formally organise the Italian Electronic Group applied to Medicine and Biology in affiliation with the International Federation of Medical and Biological Engineering (IFMBE).

Later in 1967, the constitution of the Italian Society was approved with the name "Associazione Italiana di Ingegneria Medica e Biologica" (AIIMB), with its legal home at the Engineering Faculty of University of Naples, "FEDERICO II".

The Society started with 76 members comprising medical engineers, physicists, and physicians from all over Italy. Its objectives are scientific information, education, and diffusion of knowledge with the specific aims of encouraging research, spreading scientific information, applying knowledge, and promoting collaboration with other scientific and technical groups at the national and international levels.

Registration at MEDICON 2004, Ischia, Italy

At present, the AIIMB coordinates it efforts for about 130 active Members. Most of them work at universities or scientific institutions, while others are medical and clinical engineers and physicians, working at hospitals, and other healthcare departments and industries. There are also independent medical doctors.

The AIIMB has two Working Groups, formed since 1983, on "Clinical Engineering" and on "Biomechanics". Recently a new Working Group on "Cellular Engineering" has been formed.

The AIIMB cooperates with different Italian Associations: Gruppo Italiano di Ultrasonologia Vascolare-GIUV; Società Italiana Reti Neuroniche; Associazione Italiana di Fisica Biomedica-AIF; Sezione Italiana dell' International Society of Electrophysiological Kinesiology-ISEK.

AIIMB also coordinates and cooperates with other Italian and International Technical and general Scientific Organizations and specialized activities in the field of Biomedical and Clinical Engineering within the structures of the National Health Services.

Between 1985 to 1988, AIIMB, under the direction of Prof. Marcello Bracale, Chairman of IFMBE Developing Countries Committee, coordianted a program of Clinical Engineering in many African countries. Prof. Marcello Bracale, Secretary General of AIIMB, spent time in Ethiopia, Yemen, and Kenya to organize and direct the International and Technical Course on the Management and Maintenance of Medical Equipments sponsored by WHO, IFMBE, and DANIDA. AIIMB sponsored and supported technical and industrial facilities with technical documentation for the implementation of laboratories for practical studies for students coming form different Africa Countries.

IFMBE booth at MEDICON 2004, Ischia, Italy

Presentation of Awards to IFMBE
Young Investigator Competition winners, Ischia, Italy

AIIMB participates in the Italian Group Sub-Committee 62D of the International Electrotechnical Commission for the standardization of medical equipment and in the Ministry of Health for cardiac stimulation

AIIMB, in cooperation with Italian National Engineers Register, organized different round tables and conferences on the training and the professional mobility of Biomedical Engineers in Italy and in Europe.

The Association has participated and organised many scientific activities, as well as national and international conferences. The most important are:

- In 1970, Naples, the First Italian Conference on Bioengineering.
- For many years until 1995, AIIMB organized Special Sessions of Biomedical Engineering at the International Conference of Electronics and Telecommunication, Rome, Italy.
- AIIMB, jointly with the Polish Academy of Sciences and Polish IFMBE Member Society, organized several Italian/Polish Symposia on Biomedical Engineering (Ischia, Capri, Warsaw).
- In 1977, AIIMB was invited by the president of IFMBE, Prof. Lester Goodman, and the Administrative Council of the Federation to iniate IFMBE Regional Conferences. AIIMB in cooperation with French, Israeli, Greek, Spanish, and Yugoslavian Member Societies started the MEDICON series, with the First Mediterranean Conference on Medical and Biological Engineering, Sorrento, Italy, 12-17September 1977.

AIIMB has been the Organizing IFMBE Member Society for:

- The IV Mediterranean Conference on Medical and Biological Engineering, Capri, Italy, 5-10 July 1992, and the X MEDICON AND HEALTH TELEMATICS 2004, Ischia, Naples, Italy, 31July-5 August 2004.
- The Congress on Health Telematics '95: Health in the Information Society, Ischia, Italy, 2-6 July 1995, in cooperation with the DG XIII° of European Commission EU.

Others important activities:

- CATANIA 99, AIIMB sponsored the International Conference and Exhibition of supported by the DG XIII of European Commission, Catania, Italy.
- National Congress and Exhibition TIMED, Telemedicine Information Technology and Telecommunication for Health Care, Genova, Italy, 1999, 2000, and 2001.
- National Workshop on Biomechanic Simulation in the Automotive Field, Naples, Italy, 30 April 03.
- Forum and Exhibition on Biotechnology and Bioengineering, BIONOVA 2003, Padova, Italy, 4 May 2003.
- At the Annual Italian Workshop in Bressanone, Scuola annuale di Bioingegneria, Bressanone, Italy, from September 1996 to 2008; AIIMB sponsored the Degree award "Giuseppe Francini".
- International Congress, 3rd International Workshop on Models and Analysis of Vocal Emissions for Biomedical Applications, Firenze, Italy, 2003, 2004, 2005, and 2006
- Forum and Exhibition on Biotechnology and Bioengineering, BIONOVA 2006, Padova, Italy, 6 May 06.
- V International Workshop on Models and Analysis of Vocal Emissions for Biomedical Applications, MAVEBA 2007, Firenze, Italy, 13-15 December 2007.

Prof. Marcello Bracale,
Secretary General

AIIMB "Associazione Italiana di Ingegneria Medica e Biologica"

Presidente: **Prof. Antonio Pedotti**
Secretary General: **Prof. Marcello Bracale**

Nel 1965, un gruppo di Ingegneri Elettronici, Automatici e di Medici, guidato dal Prof. G. Francini si sono formalmente riuniti per organizzare il Gruppo Italiano di Elettronica applicata alla Medicina ed alla Biologia, in affiliazione alla Federazione Internazionale di Ingegneria Medica e Biologica (IFMBE).

Nel 1967, la costituzione della Società Italiana è stata approvata con il nome di "Associazione Italiana di Ingegneria Medica e Biologica" (AIIMB) con sede legale presso la Facoltà di Ingegneria dell'Università degli Studi di Napoli "Federico II".

L'AIIMB al suo esordio contava 76 membri, tra Ingegneri Elettronici, Automatici, Fisici e Medici da ogni parte d'Italia. I suoi obiettivi sono la divulgazione e l'istruzione a carattere scientifico, con la specifica finalità di promuovere ricerche, diffusione delle informazioni scientifiche, applicazione di conoscenze e promozione di collaborazioni con altri gruppi scientifici e tecnici a livello nazionale e internazionale.

Attualmente la A.I.I.M.B. coordina circa 130 membri attivi.

La maggior parte di loro lavorano presso Università o Istituti Scientifici, mentre altri sono Medici, Ingegneri Clinici e Biomedici, che lavorano in ospedali, enti sanitari ed industrie del settore.

L'A.I.I.M.B. ha formato due gruppi di lavoro dal 1983 sull' "Ingegneria Clinica" e sulla "Biomeccanica".

Recentemente è stato costituito un gruppo di lavoro sulla "Ingegneria Cellulare".

Dal Gruppo dell'Ingegneria è gemmato l'Associazione Italiana Ingegneria Clinica AIIC

L'A.I.I.M.B. coopera con diverse altre realtà associative italiane, come ad esempio: il Gruppo Italiano di Ultrasonologia Vascolare – GIUV, la Società Italiana Reti Neuroniche, l'Associazione Italiana di Fisica Biomedica – AIF, la Sezione Italiana dell'International Society of Electrophysiological Kinesiology – ISEK.

L'A.I.I.M.B. coordina attività generali e specializzate nel settore biomedico e dell'Ingegneria Clinica, all'interno delle strutture del Servizio Sanitario Nazionale, in cooperazione con altre realtà italiane e internazionali tecnico-scientifiche.

Dal 1985 al 1988, l'A.I.I.M.B., sotto la direzione del Prof. Marcello Bracale, Chairman del Gruppo di Lavoro IFMBE per i paesi in via di sviluppo, ha collaborato ad un programma di Ingegneria Clinica per molti paesi africani. Il Prof. Marcello Bracale, Segretario Generale del AIIMB, ha così trascorso un periodo in Etiopia, Yemen e Kenya per organizzare e dirigere il Corso Tecnico Internazionale relativo alla Gestione e Manutenzione dei Servizi per la Medicina, sponsorizzato dalla OMS, IFMBE, DANIDA. L'AIIMB ha sponsorizzato e e fornito, con la collaborazione di industrie del settore,attrezzature tecniche e nutrita documentazione tecnica per l'organizzazione di lavoro pratico in laboratori ad uso di studenti provenienti da numerosi Paesi Africani.

L'A.I.I.M.B. partecipa al Gruppo Italiano Sub-Comitato 62D della Commissione Elettrotecnica Internazionale per la standardizzazione delle attrezzature mediche e ha collabora con il Ministero della Salute per studi sulla stimolazione cardiaca.

L'AIIMB in collaborazione con l'Ordine Nazionale Italiano degli Ingegneri ha organizzato diverse tavole rotonde e conferenze sulla formazione e la mobilità professionale degli ingegneri biomedici in Italia e in Europa.

L'Associazione ha partecipato ed organizzato numerose attività scientifiche e congressi nazionali ed internazionali. I più importanti sono:

- nel 1970, in Napoli: la Prima Conferenza Italiana sulla Bioingegneria;
- per molti anni, fino al 1995, l 'AIIMB è stata organizzatrice di Sessioni Speciali inerenti l'Ingegneria Biomedica, svoltesi durante il Congresso annuale Internazionali di Elettronica e Telecomunicazioni - Roma, Italia;
- l' AIIMB in collaborazione con l'Accademia Polacca delle Scienze e la Associazione Polacca affiliata all' IFMBE, ha organizzato numerosi simposi italo-polacchi sul tema dell'Ingegneria Biomedica (Ischia, Capri, Varsavia);
- nel 1977, l' AIIMB, su invito del Presidente del IFMBE, Prof. Lester Goodman e del Consiglio Direttivo della Federazione, ha avviato l'attività delle Conferenze Regionali dell' IFMBE. L'AIIMB in collaborazione con le associazioni francese, israeliana, greca, spagnola e jugoslava ha dato inizio alla triennale Conferenza Internazionale MEDICON, con la prima edizione in Sorrento, Italia, dal 12 al 17 settembre 1977.

L'AIIMB è stato organizzatore per conto dell' IFMBE:

- IV MEDICON - Capri, Italia 5-10 luglio 1992;
- X MEDICON and "Health e Telematics" 2004 – Ischia, Napoli, Italia 31 luglio - 5 agosto 2004;
- Congresso sulla "Health & Telematics '95: Health in the Information Society" - Ischia, Italia. 2-6 luglio 1995, in collaborazione con la DG XIII della Commissione Europea UE.

Altre importanti attività sono state dedicate:

- "CATANIA 99": l' AIIMB ha promosso la Conferenza e Mostre Internazionale, con il sostegno della DG XIII° della Commissione Europea, Catania, Italia;
- Congresso Nazionale e Mostra TIMED "Telemedicine Information Technology and Telecommunication for Health Care", Genova, Italia, anni 1999, 2000 e 2001;
- Seminario Nazionale "Biomechanic Simulation in the Automotive Field", Napoli, Italia 30 aprile 2003;
- Forum e Mostra sulle Biotecnologie e la Bioingegneria - BIONOVA 2003 – Padova, Italia, 4 maggio 2003;
- alla "Scuola Annuale di Bioingegneria", Bressanone, Italia, dal settembre 1996 al 2008, l' AIIMB sponsorizza il premio di laurea "Giuseppe Francini";

- Congresso Internazionale “3rd International Workshop on Models and Analysis of Vocal Emissions for Biomedical Applications”, Firenze, Italia, anno 2003, 2004, 2005 e 2006;
- Forum e Mostra sulle Biotecnologie e la Bioingegneria - BIONOVA 2006 – Padova, Italia, 6 maggio 2006;
- V workshop internazionale “Models and Analysis of Vocal Emissions for Biomedical Applications - MAVEBA 2007”, Firenze, Italia, 13 - 15 dicembre 2007

Prof. Marcello Bracale
Secretary General

The Italian Association of Clinical Engineers (AIIC)

Clinical Engineers are healthcare professionals focused on safe, appropriate, and economical management of technologies in health services, working at different levels in public and private hospital enterprises, IRCCS (Scientific Institute for Research, Hospitalization and Health Care), Services Enterprises, etc.

The most important activities of Clinical Engineers are: evaluation of biomedical technologies and medical devices with *Health Technology Assessment* (HTA) methodologies; safety and quality testing; training of equipment users, maintenance management, information technology, and risk management.

The Italian Association of Clinical Engineers (AIIC) has been working in Italy since 1994, with the aim of protecting the professional stature of clinical engineers and promoting Clinical Engineering Departments in the Healthcare Enterprises as part of both the routine management and maintenance of technological resources and the governance of technology innovation (in terms of safety, efficacy, and efficiency), particularly concerning biomedical equipment and medical devices which are absolutely indispensable for providing healthcare services.

Furthermore, the Association mission is to promote the scientific research and the technological knowledge in the field of Clinical Engineering together with the dissemination of global information related to Clinical Engineering Departments and applied to heath technologies management within healthcare organizations.

The number of AIIC members has grown rapidly in the last few years (about 400 of today members vs. 100 members in 2004), as well as has its collaboration with major Institutions, first of all, the National Health Service and Minister of Health (Technical Committee on Clinical Risk, Working Group for Patient Safety, Working Group on Technological Systems in Health); AIIC is present in several Technical Committees of CEI (Italian Electrotechnical Committee - Comitato Elettrotecnico Italiano), the Italian institution, formally recognized by the Italian Government and by the European Union, that develops and publishes technical standards in the electrical, electronic, and telecommunication fields. AIIC is also present in the Integrating Healthcare Enterprise (IHE) Committee and has offered to contribute to INAIL's (National Institute for the Injuries in the Workplace) document "Hospital safety: tools for risk assessment and management" ("La Sicurezza in Ospedale: strumenti di valutazione e gestione del rischio"), particularly regarding biomedical equipment related risks.

Although the Clinical Engineers have been working in the Italian Health organization for more than thirty years, only in the few last months has Italian legislation taken its first steps toward institutional recognition to the Clinical Engineering Departments with a Government draft law, that is still in discussion in the Italian Parliament: AIIC expressed its availability to the Minister of Health to support the development of this important program.

Associazione Italiana Ingegneri Clinici

L'*Associazione Italiana Ingegneri Clinici* è stata fondata nel 1993 con lo scopo istituzionale di tutelare la figura professionale dell'Ingegnere Clinico contribuendo a diffondere i Servizi di Ingegneria Clinica all'interno delle aziende sanitarie. *AIIC* riunisce a livello nazionale professionisti impegnati nell'ambito dell'Ingegneria Clinica a diversi livelli: Strutture Sanitarie Ospedaliere pubbliche e private, Istituzioni centrali/regionali, società di servizi, ecc. L'Ingegnere Clinico partecipa alla cura della salute garantendo un uso sicuro, appropriato ed economico della strumentazione e delle attrezzature biomedicali ed *info - telematiche clinico - assistenziali* presenti nei servizi socio-sanitari esercitando, tra le altre, le seguenti attività:

- *valutazione di tecnologie sanitarie e sistemi sanitari con le metodologie dell'Health Technology Assessment;*
- *programmazione degli acquisti di tecnologie;*
- *gestione delle tecnologie e progettazione funzionale;*
- *collaudi di accettazione;*
- *gestione della manutenzione e delle attività conseguenti;*
- *gestione della sicurezza delle tecnologie;*
- *formazione sull'utilizzo delle tecnologie;*
- *integrazione delle tecnologie nell'ambiente ospedaliero;*
- *informatica clinica ed "Information Technology";*
- *sviluppo di software, procedure e dispositivi medici.*

L'*Associazione* è fortemente cresciuta nell'ultimo decennio non solo in termini di iscritti (oggi circa 400, raddoppiati rispetto a 4 anni fa) ma anche sotto il profilo della visibilità esterna e del coinvolgimento attivo in *Enti* ed *Istituzioni*, come testimoniato dalle numerose attività nelle quali *AIIC* è chiamata a dare il proprio contributo specialistico (es. Ministro della Salute: *Commissione Tecnica sul Rischio Clinico,* GdL per la *Sicurezza dei pazienti,* GdL sui *Sistemi Tecnologici in Sanità,* GdL *Progettazione di strumenti per il miglioramento della qualità con riferimento all'ottimizzazione della logistica e dei percorsi intra-ospedalieri, Nucleo di Valutazione e Verifica degli investimenti pubblici in sanità*; INAIL: GdL *"La Sicurezza in Ospedale: strumenti di valutazione e gestione del rischio";* Age.na.s.: progetto *"Centro di Osservazione delle Tecnologie biomediche Emergenti";* CEI-*Comitato Elettrotecnico Italiano ed* IHE-*Integrating Healthcare Enterprise;* società scientifiche mediche; associazioni professionali in sanità). Gli obiettivi futuri sono finalizzati al rafforzamento dei contatti istituzionali, delle *partnership* con il mondo accademico (*GNB – Gruppo Nazionale di Bioingegneria*) e con i colleghi esteri (cfr. affiliazione con *IFMBE - International Federation for Medical and Biological*

Engineering, *partnership* con *ACCE – American College of Clinical Engineering*) ed in generale alla promozione dell'immagine e della formazione degli Ingegneri Clinici.

AIIC ha espresso recentemente la sua integrale disponibilità al *Ministero del Lavoro, della Salute e delle Politiche Sociali* a partecipare ai tavoli di programmazione ministeriale nell'ambito dell'innovazione tecnologica in sanità ed a supportare la declinazione del recente *DdL "Disposizioni per la tutela del paziente e per la riparazione, la prevenzione e la riduzione dei danni derivanti da attività sanitaria"*, relativamente all'istituzionalizzazione dei *Servizi di Ingegneria Clinica* (cfr. *Art. 8*).

The Japanese Society for Medical and Biological Engineering (JSMBE)

The Japanese Society for Medical and Biological Engineering (JSMBE) was established on November 10, 1962. JSMBE has 2810 members and 18 corporate members as of January 15, 2009. The current president is Dr. Hideo Kusuoka (Director, National Hospital Organization Osaka Hospital). JSMBE promotes academic activities in medical and biological engineering both for the application of engineering methodology (electrical, electronic, mechanical, chemical material, information engineering, and so on) to medical science, clinical science and biology, and the application of medical and biological knowledge to engineering. JSMBE seeks to contribute to the health and welfare of the society through its academic activities.

JSBME organizes an annual academic conference having approximately 1000 participants. It also organizes a Medical and Biological Engineering Symposium for stimulating intense discussion among young researchers in this field. JSMBE has 8 local divisions in Japan (Hokkaido, Tohoku, Koshinetsu, Hokuriku, Kansai, Chugoku-Shikoku, and Kyushu division respectively). They organize their own symposiums, conferences, and seminars to promote academic activities in each district. JSBME cooperates with more than 80 conferences and seminars organized by Japanese and International Societies. JSMBE has 23 specialized study groups on individual specific topics.

JSMBE sponsors several awards to recognize excellent papers, research activities, and industrial contributions. It also has the Ogino Award to provide a grant for promising research.

In the realm of education for Medical and Biological Engineering in Japan, JSMBE organizes 5 training courses a year for clinical engineers and conducts certification examinations for clinical engineers, with approximately 5000 applicants. It also provides accreditation service for clinical medical engineering specialists. JSMBE has accredited 444 specialists in total.

JSMBE issues *Transactions of the Japanese Society for Medical and Biological Engineering* quarterly including review articles in the related areas and approximately 40 original peer reviewed papers. It also published the *Medical Engineering Dictionary* and textbook series in Medical and Biological Engineering. Some of the textbooks are now being translated in Asian countries.

JSBME has actively collaborated with IFMBE and has been involved with various IFMBE's activities. Since 2006, IFMBE, JSMBE, and related Asian societies established IFMBE Asia-Pacific Traveling Fellowships for Junior Faculty in Biomedical Engineering. It supports mutual visits of young Asian researchers in this field in Asian districts and their participation in IFMBE World Congress.

JSMBE actively contributes to establish engineering standards for Medical devices and systems in Japan including nomenclature of medical devices and systems.

Participants of APCMBE 2005, Tsukuba, Japan

日本生体医工学会 (Japanese Society for Medical and Biological Engineering)

日本生体医工学会は，1962年11月10日設立に「日本エム・イー学会」として設立され，2005年４月１日に現在の「日本生体医工学会」の名称に変更した．
現在の会員数は2009年1月15日現在正会員2810名，法人会員18社である．現在の会長は楠岡英雄先生（大阪医療センター院長）である．本学会は，医学，生物学における電子工学，機械工学，化学工学，材料工学，情報工学等の工学的な方法，及び工学における医学，生物学的知見の応用に関する研究の発展，知識の交流及び社会における事業の振興をはかることを目的とし，以下の事業を行っている．

Mingling during the Welcome Reception

年１回約１０００名の参加者による定期大会を開催している．また若手研究者の論文発表の場として年一度生体医工学シンポジウムを開催している．本学会には北海道支部，東北支部，甲信越支部，東海支部，北陸支部，関西支部，中四国支部，九州支部の８つの地方支部が組織されており，各地方における講演会などの独自の活動を推進している．また，２３の専門別研究会が組織されており，生体医工学に関する個別課題について学術推進活動を行っている．

Poster session at the Asian Pacific Conference on
Medical and Biological Engineering, Tsukuba 2005

また教育に関連して臨床工学技術の高度化推進，臨床工学技師の教育のための，年５回の講習会を開催し，約５００名以上の参加を得ている．また関連し実力検定試験（第一種：受験者数約4500名，第二種：受験者数役400名）ならびに臨床ME専門認定制度による認定活動を実施している．

学会誌「生体医工学（Transactions of the Japanese Society for Medical and Biological Engineering） を年４回発刊し，解説記事とともに年間約４０件の原著論文を掲載している．また生体医工学関連の用語集・辞典の編纂，教科書の編集発行を行い．一部についてはアジア諸国において外国語訳化も行われている．

国際医用生体工学連合への加盟し積極的な連携活動を実施している．2006年より" IFMBE Asia-Pacific Traveling Fellowships for Junior Faculty in Biomedical EngineeringをIFMBEならびにアジア関連学会と協力して総説し，IFMBE若手研究者によるアジア地域内の研究施設内の相互訪問ならびにIFMBE World Congress 派遣への支援活動を行っている.

また，関連する内外の関連諸学協会との連絡並びに協力活動を行い年間８０件程度の協賛事業を行っている.

公的な活動としては，わが国における関連規格制定に関する協力活動及び用語の統一に関する活動に寄与している.

日本生体医工学会は日本生体医工学会論文賞，研究奨励賞，新技術開発賞，研究助成としての荻野賞を有し，優秀な研究者，開発者の顕彰ならびに研究振興を図っている.

Presentation of Awards to IFMBE
Young Investigator Competition winners, 2005

Korea Society of Medical and Biological Engineering (KOSOMBE)

대한의용생체공학회
The Korean Society of Medical & Biological Engineering

The Korea Society of Medical and Biological Engineering (KOSOMBE) celebrated its 30th anniversary on November 13, 2009. Founded in 1979 by several Korean BME pioneers, it has grown to one of the largest and most productive academic societies in Korea. With about 600 active regular members and 1500 student members, the total number of registered members reached 3600 this year. Active members joined the society from about 40 academic institutes offering BS, MS, and PhD degrees in BME, 25 companies in health industry, and several research institutes. Among seventeen in total, eleven former presidents have joined the commemorative ceremonies during annual conferences in May and November, 2009.

Preparing for the WC 2006 Welcome Reception

Even though some of them had health problems, they were eager to share with the rising generation their memories and experiences of the rapid growth of the society and the BME area itself for the last 30 years. The current president, Professor Sun Il Kim announced several enthusiastic plans for the next 30 years. The KOSOMBE will start publishing a new English journal called Biomedical Engineering Letters. BME researchers and students from all over the world will find it easy to access papers published in this second official journal of the KOSOMBE in addition to the existing Journal of Biomedical Engineering Research. The society took the initiative on concerted action to reform undergraduate BME curriculum for the next generation of the bio era. Joint conferences with other societies were proposed and are planned to accelerate academic fusion among related disciplines. A task force is formed within the society to publish a white book about BME research in Korea. The next president of the KOSOMBE, Professor Do-Un Jeong will be in charge the society for the coming two years of 2010 and 2011. Council members elected Professor Jin Ho Cho as the President-elect for the year of 2012.

WC 2006 Opening Ceremony
L-R: Joachim Nagel, IFMBE President, Barry Allen, IUPESM President and World Congress Chair, Kim Sun.

The Korean Society of Medical & Biological Engineering

대한의용생체공학회
The Korean Society of Medical & Biological Engineering

2009년 11월 13일, 대한의용생체공학회의 창립 30주년을기념하였다. 대한의용생체공학회는 1979년 몇몇 한국 의공학의 선구자들이 설립하여 한국에서 매우 활성화된 학회로 성장하였다. 올해까지 600여명의 정회원과 1500여명의 학생회원을 비롯하여 총 3600여명이 회원으로 등록하였다. 의공학 학위과정을 개설한 40여개의 대학과 25개의 의료기 관련 업체 그리고 몇몇 연구기관들이 참여하여 회원으로서 활동하고 있다. 이제까지의 17명의 전임 회장 중 11명이 2009년 5월과 11월에 열린 창립 기념 학술대회에 참석하였다. 그들 중 몇몇은 건강이 좋지 않음에도 불구하고 지난 30년 동안 급속하게 성장한 학회와 의공학 분야에 대해 다음 세대들과 추억과 경험을 나누고자 하였다.

현 회장이신 김선일 교수는 다가올 30년에 대하여 몇 가지 주요 계획을 공표하였다. 먼저 의공학회는 "Biomedical Engineering Letters"라 불리는 새로운 영문판 학회지를 발행할 것이고, 그럼으로써 많은 의공학 연구자와 학생들은 기존의 의공학회지(Journal of Biomedical Engineering Research) 뿐만 아니라 새로이 만들어질 학회지(Biomedical Engineering Letters)를 통해서 많은 논문을 쉽게 접할 수 있게 될 것이다. 다음으로 학회는 솔선해서 의공학 분야의 다음 세대를 위하여 대학의 의공학 교육과정을 재정립하기로 하였다. 또한, 다른 학회와 교류를 통하여 관련 학문분야 간의 학문적 융합 및 가속화를 계획하였으며, 학회에서는 의공학 연구에 관한 백서를 발행하기 위해 특별위원회를 발족하였다. 대한의용생체공학회의 다음 회장이신 정도언 교수는 2010과 2011년, 2년간 회장직을 맡을 것이고, 이사회는 또한 조진호 교수를 2012년도의 차기회장으로 선출하였다.

The Latvian Medical Engineering and Physics Society (LMEPS)

The Latvian Medical Engineering and Physics society (LMEPS) currently counts 35 professionals w orking in the field of biomedical engineering, medical physics, as well as adjacent fields. The goal of the Society is to promote development of the biomedical engineering and medical physics in Latvia.

In the year 2008, LMIFB, in cooperation with professional associations, hospitals, and educational institutions (Latvian Radiologists Association, Latvian Oncology Center, Riga Technical University, P. Stradina University Hospital, and University of Latvia), developed national Professional Standard for the Medical Physicist. Development of such a standard becomes an important milestone toward certification of medical physicists at the national level.

LMIFB, in cooperation with the Latvian Radiologists Association and Riga Technical University, organize regular courses on Radiation Protection for healthcare professionals. Up to 40 participants annually attend the courses.

At the international level, LMIFB, together with Riga Technical University and the University of Latvia, hosted the 14th Nordic-Baltic Conference on Biomedical Engineering and Medical Physics, triennial regional conference of IFMBE. The conference took place on June 16-20, 2008, in Riga, capital city of Latvia. The meeting brought together about 240 participants from more than 30 countries and played an important role for the popularization and development of biomedical engineering in the Baltic region.

Conference board

Latvijas medicīnas inženierzinātņu un fizikas biedrība

LMIFB

Latvijas medicīnas inženierzinātņu un fizikas biedrība (LMIFB) uz doto brīdi apvieno 35 profesionāļus, kas strādā medicīnas inženierzinātnes un medicīnas fizikas virzienā, ka arī blakusvirzienos. Biedrības mērķis ir sekmēt biomedicīnas inženeirznātnes un medicīnas fizikas attīstību Latvijā.

2008. gadā LMEPS, kopā ar speciālistiem no profesionālām asociācijām, slimnīcām un izglītības iestādēm (Latvijas Radiologu Asociācija, Latvijas Onkoloģijas centrs, Latvijas Universitāte, Rīgas Tehniskā universitāte, P. Stradiņa Universitātes klīniskā slimnīca) izstrādāja medicīnas fiziķa nacionālā profesionālā standarta projektu. Šis darbs ir svarīgs posms ceļā līdz medicīnas fiziķa nacionālās sertifikācijas sistēmai.

LMIFB, sadarbībā ar Latvijas Radiologu Asociāciju un Rīgas tehnisko universitāti regulāri organizē Radiācijas drošības kursus veselības aprūpes speciālistiem. Kursus apmeklē vairāk par 40 klausītāju gadā.

Starptautiskā līmenī LMEPS, kopā ar Latvijas Universitāti un Rīgas Tehnisko universitāti 2008. g 16 – 20. jūnijā organizēja 14 Ziemeļu –Baltijas Biomedicīnas inženierzinātnes un fizikas konferenci, kas ir Starptautiskās Medicīnas un Bioloģijas Inženierzinātnes federācijas reģionāla konference. Tajā piedalījās ap 240 zinātnieku vairāk kā no 30 valstīm. Konference deva nozīmīgu ieguldījumu biomedicīnas inženierzinātņu popularizācijai un attīstībai Baltijas valstīs.

Participants gather at the NBC2008 venue entrance

Malaysian Society of Medical and Biological Engineering (MSMBE)

Membership drive at University Malaysia perlis on March 28th 2008, Fatimah Ibrahim the MSMBE's secretary second from left, Wan Abu Bakar Wan Abas, the MSMBE's president 3rd from right

The first event organized by the Society was the 3rd Kuala Lumpur International Conference on Biomedical Engineering (Biomed 2006) on December 11-14th, 2006. The conference was also co-organized by the Department of Biomedical Engineering, University of Malaya; and Inje University, Korea. The conference was endorsed by the IUPESM and IFMBE, along with a number of other local organizations.

First front row from left Wan Abu Bakar Wan Abas, President, MSMBE
3rd front row from left Mikoto Makoto Kikuchi, President, IFMBE
2nd front row from right, Fatimah Ibrahim, MSMBE's secretary

The Mexican Biomedical Engineering Society (SOMIB)

Biomedical Engineering in Mexico (A Survey)

Solorzano J, Cadena M, Infante O, Prieto F, Suaste E and Martinez A
Present and former presidents of SOMIB, www.somib.org.mx

BACKGROUND

The practice of medicine and surgery by pre-Columbian civilizations in Mexico has been extensively documented.[1]

An interesting pictorial history of cardiology, that is also a beautiful reference to the history of biomedical engineering, can be admired in a mural at the National Institute of Cardiology, painted by Diego Rivera in 1947. Many of the historical milestones there depicted took place in Mexico or in Latin America.[2]

HISTORICAL OVERVIEW

No one can tell exactly when modern Biomedical Engineering began to be practiced in Mexico. Our mention of the National Institute of Cardiology is not coincidental. Thanks to the vision of its founder Dr. Ignacio Chavez, the Institute has hosted many renowned scholars whose contribution to science and technology cannot be overestimated. One was Arturo Rosenblueth (1900-1970), whose name is associated with the development by Norbert Wiener's concept of cybernetics, and, who obtained from Albert Grass, the instruments he needed for research in neurophysiology often giving ideas for their design and improvement.[3] Research and development related to technological applications in cardiology today is the task of the Instrumentation Department at the National Institute of Cardiology.

Rosenblueth would be a co-founder, in 1961, of the Center for Research and Advanced Studies (CINVESTAV) of the National Polytechnical Institute (IPN).[4] A few years later, 1971, Joaquin Remolina founded CINVESTAV, the rather unlikely Bio-Electronics Section, within the Department of Pharmacology. A physician turned designer and builder of electronic devices, Joaquin was the father of a prolific biomedical engineering tradition that has disseminated and thrived in multiple institutions and enterprises.[5]

The National Autonomous University of Mexico (UNAM) has also been the cradle of achievements in biomedical engineering. It would take too long to make a list of the well known and dearly remembered scientists and technologists who have labored and taught in the laboratories, workshops, and classrooms of UNAM.[6,7]

PRESENT PANORAMA

Dozens of professional institutions offer programs in Biomedical Engineering in Mexico. The earliest programs for undergraduate studies in biomedical engineering started almost simultaneously in the Autonomous Metropolitan University (UAM) and Universidad Iberoamericana (UIA), between 1974 and 1976.[6] Since that time, undergraduate and postgraduate programs followed in several universities and technological institutions across the country. Hundreds of biomedical engineers have completed their baccalaureate, master's, or doctoral degree in Mexican institutions.

The Mexican Biomedical Engineering Society (SOMIB)[8] was founded in 1978. The Society now comprises the majority of professionals and academics engaged in biomedical engineering in the country. It organizes a yearly congress and publishes an internationally registered journal. SOMIB is a member of IFMBE and CORAL (the regional biomedical engineering committee for Latin America). There is a Mexican Chapter of IEEE-EMBS. Research papers on Biomedical Engineering subjects are published by Mexican authors in influential periodicals al home and abroad.

Most Mexican hospitals have a biomedical engineering department. Much remains to be done to make the transition from departments that are dedicated solely to servicing equipment to full-fledged clinical engineering departments practicing all aspects of medical device management. There is also much to be done in integrating hospital practice with academic research and development. The Mexican Institute for Social Security (IMSS)[9], which is the main provider of medical services in the country, has technical departments in its hospitals and medical centers, although under names different to Biomedical Engineering and with only indirect relation with SOMIB.

The Ministry of Health has fostered, for over a decade, a law office called the National Center for Technological Excellence (CENETEC).[10] This center is destined to play an important role in the development of national standards for medical devices and technology.

The medical device industry and market have been slow to develop in Mexico. Nevertheless, this branch of business has grown steadily and there is considerable activity in importing, marketing, and technical services both by Mexican firms and representatives of foreign companies.

ACKNOWLEDGMENT

We would like to acknowledge the contributions that dozens of people, both alive and dead, have made to the biomedical engineering profession in Mexico. From classrooms, research and instrumentation laboratories, hospital engineering departments, and private industry, all are certainly key figures in the past, present, and future of biomedical engineering.

References

(For ease of access, we have given preference, in this short communication, to references from Internet sites and documents).

1. http://www.nlm.nih.gov/hmd/pdf/precolumbian.pdf
2. http://diegorivera.com/index.php
3. http://eng.wikipedia.org/wiki/Arturo_Rosenblueth/
4. www.cinvestav.mx/
5. Zottoli, SJ: The Origins of the Grass Foundation. Biol. Bull. 2001(Oct): 201; 218 -226. http://www.grassfoundation.org/about/pdf/history.pdf
6. http://web.mac.com/ci3m/Web/Site%203/Material%20Adicional_files/IngBiomed1.pdf
7. http://dctrl.fi-b.unam.mx/~capituloestudiantil/faq.php
8. www.somib.org.mx/
9. www.imss.gob.mx/
10. www.cenetec.salud.gob.mx/

Ingeniería Biomédica en México (Una Visión Sucinta)

Solorzano J, Cadena M, Infante O, Prieto F, Suaste E and Martinez A
Presidentes actual y anteriores de la SOMIB, www.somib.org.mx

ANTECEDENTES

La práctica de la medicina y la medicina por las civilizaciones precolombinas de México ha sido extensamente documentada. [1]

Una representación pictórica interesante de la historia de la cardiología que es al mismo tiempo una bella referencia a la historia de la ingeniería biomédica, puede ser admirada en el mural que pintó Diego Rivera en 1947 en el Instituto Nacional de Cardiología. [2]

VISIÓN HISTORICA

Nadie puede decir con exactitud cuando comenzó a practicarse en México la ingeniería biomédica moderna. Nuestra mención del Instituto Nacional de Cardiología no es casual. Gracias a la visión de su fundador, el doctor Ignacio Chávez, el instituto ha alojado muchos académicos de renombre, cuya contribución a la ciencia y la tecnología no pueden ser sobreestimadas. Uno de ellos fue Arturo Rosenblueth (1900 – 1970) cuyo nombre está asociado con el desarrollo por Norbert Wiener del concepto de cibernética y quien obtuvo de Albert Grass los instrumentos necesarios para la investigación neurofisiológica y contribuyó a ello con ideas para su diseño y mejoría. [3] La investigación y el desarrollo de aplicaciones tecnológicas en cardiología es hoy en día la labor del Departamento de Instrumentación del Instituto Nacional de Cardiología.

Rosenblueth habría de ser co-fundador, en 1961, del Centro de Investigación y Estudios Avanzados (CINVESTAV) del Instituto Politécnico Nacional (IPN). [4] Pocos años más tarde (1971), Joaquín Remolina fundó en el CVINVESTAV la improbable Sección de Bioelectrónica, dentro del Departamento de Farmacología. Médico transformado en notable diseñador y constructor de implementos electrónicos, Joaquín sería el padre de una tradición en ingeniería biomédica que se ha diseminado y proliferado en múltiples instituciones y empresas. [5]

También la Universidad Nacional Autónoma de México (UNAM) ha sido cuna de realizaciones en ingeniería biomédica. Sería demasiado largo tan sólo el hacer una lista de científicos y tecnólogos ampliamente conocidos que han laborado y enseñado en los laboratorios, talleres y aulas de la UNAM. [6, 7]

PANORAMA PRESENTE

Decenas de instituciones profesionales ofrecen programas en ingeniería biomédica en México. Los primeros programas de pregrado en ingeniería biomédica comenzaron de manera casi simultánea en la Universidad Autónoma Metropolitana (UAM) y en la Universidad Iberoamericana (UIA), alrededor de los años 1974 a 1976. [6] A partir de esas fechas los programas de pregrado y de posgrado se sucedieron en diversas universidades e institutos tecnológicos a lo largo y ancho del país. Cientos de ingenieros biomédicos han completado su licenciatura, maestría o doctorado en instituciones mexicanas.

La Sociedad Mexicana de Ingeniería Biomédica (SOMIB) [8] fue fundada en 19778. La sociedad agrupa ahora a la mayor parte de los profesionales y académicos que se dedican a la ingeniería biomédica en el país. Organiza un congreso cada año y publica una revista con registro internacional. La SOMIB es miembro de la IFMBE y de CORAL (el comité regional de ingeniería biomédica para la América Latina). Hay un capítulo mexicano de la Sociedad de Ingeniería en Medicina y Biología de la IEEE. Los autores mexicanos publican artículos referentes a temas de ingeniería biomédica en revistas de prestigio en el país y en el extranjero.

La mayor parte de los hospitales mexicanos tienen un departamento de ingeniería biomédica. Mucho queda aún por hacerse para dar el paso a partir de los departamentos que se dedican únicamente al servicio de los equipos hacia departamentos de ingeniería clínica que practiquen todos los aspectos de la gerencia de equipo médico. También falta mucho por hacer en cuanto a la integración de la práctica hospitalaria con la investigación y el desarrollo. El Instituto Mexicano del Seguro Social (IMSS), [9] que es el principal proveedor de servicios médicos del país, tiene departamentos técnicos en sus hospitales y centros médicos, aunque con nombres diferentes al de ingeniería biomédica y con una relación solamente indirecta con la SOMIB.

La Secretaría de Salud ha auspiciado durante más de una década una oficina normativa que recibe el nombre de Centro Nacional de Excelencia Tecnológica en Salud (CENETEC) [10] Este centro está llamado a desempeñar un papel importante en el desarrollo de los estándares nacionales para los dispositivos y la tecnología en medicina.

La industria de los equipos médicos se ha desarrollado con lentitud en México. Sin embargo esta rama de los negocios ha crecido constantemente y hay una actividad considerable en importación, comercialización y servicios técnicos a cargo de firmas mexicanas y de representantes de compañías extranjeras.

RECONOCIMIENTO

Nos gustaría rendir debido reconocimiento a los aportes que docenas de personas, tanto vivas como difuntas, han hecho a la profesión de la ingeniería biomédica en México. Desde las aulas, los laboratorios de investigación y de instrumentación y la industria privada, son todas ellas sin duda figuras clave en el pasado, presente y futuro de la ingeniería biomédica.

References

(Pensando en la facilidad del acceso a las mismas, hemos dado preferencia en esta corta comunicación a las referencias a sitios y documentos en Internet).

1. http://www.nlm.nih.gov/hmd/pdf/precolumbian.pdf
2. http://diegorivera.com/index.php
3. http://eng.wikipedia.org/wiki/Arturo_Rosenblueth/
4. www.cinvestav.mx/
5. Zottoli, SJ: The Origins of the Grass Foundation. Biol. Bull. 2001(Oct): 201; 218 -226. http://www.grassfoundation.org/about/pdf/history.pdf
6. http://web.mac.com/ci3m/Web/Site%203/Material%20Adicional_files/IngBiomed1.pdf
7. http://dctrl.fi-b.unam.mx/~capituloestudiantil/faq.php
8. www.somib.org.mx/
9. www.imss.gob.mx/
10. www.cenetec.salud.gob.mx/

Mongolian Society of Biomedical Engineering

The Mongolian Society for Biomedical Engineering (MSBME) was founded in 1999 by Prof. B. Enkhjargal, and is, today, well established at the Mongolian universities and hospitals and companies. One of the goals of the Society is to build a network for education, research, manufacturing and politics within the background of biomedical engineering (BME). With the organization of regular scientific meetings, all these fields are brought together. Annual scientific prices stimulate also younger members to play an active role in the Society.

With this in mind, the MSBME aims at establishing contacts, exchanges of information and experience, as well as cooperation between industrial and scientific people who want to apply physical and technical concepts to biology and medicine.

In 2009, MSBME included an active membership of 16 collective members and more than 70 individual members. Most of the individual members work at the universities, the universities of applied sciences, and the institutes. Other members are engineers and physicians working at hospitals and in biomedical and medical companies.

The journal “Эмнэлгийн техник, технологи/ Medical equipment and advanced technology”, which is also the official journal of the MSBME.

Club for undergraduate biomedical engineering (CUBE) is a student-run club whose mandate is to promote biomedical engineering at the undergraduate level.

Mongolian Society of Biomedical Engineering

Монголын Эмнэлгийн Инженерүүдийн Нийгэмлэгийг анх 1999 онд Проф,Б.Энхжаргал үүсгэн байгуулсан бөгөөд өнөөдөрийн байдлаар Монгол улсын эмнэлгийн инженер бэлтгэдэг их дээд сургуулиуд болон эмнэлэг, ханган нийлүүлэгч байгуулагууд хамтран ажиллаж байна. Манай нийгэмлэгийн гол зорилго бол эмнэлгийн инженерүүдийн мэргэжил дээшлүүлэлт, сургалт, эрдэм шинжилгээний ажил болон үйлдвэрлэлийг дэмжин ажиллах явдал юм. Нийгэмлэгээс эрдэм шинжилгээний хурал болон зөвлөлгөөнийг тогтмол явуулдаг ба энэ ажилд залуу инженерүүд гол үүрэг гүйцэтгэн ажилладаг юм.

2009 онд байнгын идэвхтэй гишүүн 16 ба 70 орчим гишүүнтэй үйл ажиллагаагаа явуулж байна. Монголын Эмнэлгийн Инженерүүдийн Нийгэмлэгээс "Эмнэлгийн техник, технологи" сэтгүүл гаргадаг. 2009 оноос Монголын Эмнэлгийн Инженерүүдийн Нийгэмлэгийн Тэргүүнээр УКТЭ-ийн Эмнэлгийн техникийн албаны дарга, инженер Н.Гэрэлт-Од ажиллаж байна.

Мөн Монголын Эмнэлгийн Инженерүүдийн Нийгэмлэгийн дэргэд оюутан залуучуудын "Залуу инженер" клуб ажилладаг бөгөөд Тэргүүнээр У.Батсүх ажиллаж байна.

The Netherlands' Society for Biophysics and Biomedical Engineering

Founded April 28, 1978

The Society was founded to initiate contacts between biophysicists and biomedical engineers and to promote mutual scientific discussions. Our aim is to further the development of the physical and technological aspects of medical and biological research. The Society has 4 major fields of interest: organ physics, cellular and molecular physics, neurophysics, and technical biophysics.

Biomedical Engineering (BME) was added to the name of the Society in 2000, reflecting the increased interest in BME in the Netherlands. Since 1997, BME education has been initiated at the Eindhoven University of Technology, the University Twente, and the University of Groningen. The total enrolment of students is more than 200 per year. A unique, fully integrated approach with the life sciences was incorporated from the start of the BME education program and includes several components of Design-Centered Learning, an engineering approach of Problem-Based Learning. BME is the first engineering discipline between the classical engineering disciplines and life sciences; it is anticipated that many other disciplines will develop at this interface. Present examples are Medical Engineering (Eindhoven University of Technology) and Technical Medicine (University Twente).

Internationally, the life sciences are attaining a more prominent role and at the same time become more and more dependent on contributions from the fields of biomedical engineering and biophysics. In the Netherlands, as internationally, a shift has occurred towards genetic and cellular processes. Biomedical engineering and biophysics will be crucial in unravelling the concomitant problems. That expertise will obviously also stay just as important for understanding the organ and organismal levels as well as for measurement techniques as they have been in the past decades. It is, therefore, very important that national and international grant agencies use biomedical engineering and biophysics as major keywords and incorporate this expertise in the review panels. Dutch scientists have always recognized the need for international collaboration to ensure a long term vision with a continued focus on basic research and research into quality of measurements. They have, therefore, played a role in the founding and running of various international professional organizations like the IFMBE and EAMBES as members of boards, councils, and working groups.

The Netherlands has a history of close collaboration between medical doctors, physicists, and engineers. This has led to a solid understanding of the underlying physical principles of (patho) physiological processes and the design of apparatuses to measure the relevant quantities. Current trends are that on the one hand, the expertise of "old fashioned" design knowledge and craftsmanship is disappearing, but on the other hand modelling and new rapid prototyping methods are emerging that rely less on craftsmanship and more on software and computers. Whatever the tools will be in the future, the knowledge and experience of biomedical engineering and biophysics will always play a vital role in research in the life sciences.

A.C. Linnenbank and D.W. Slaaf

De Vereniging voor Biofysica and Biomedische Technologie

Opgericht 28 April 1978

De vereniging is opgericht om contacten te leggen tussen biofysici en biomedisch ingenieurs en om onderlinge wetenschappelijke uitwisseling en discussie te bevorderen. Ook het bevorderen van de ontwikkeling van kennis op het gebied van fysica en techniek binnen het medisch en biologisch onderzoek behoort tot haar doelstellingen. De vereniging heeft vier afdelingen: orgaanfysica, cellulaire en moleculaire fysica, neurofysica en technische biofysica. Biomedische Technologie (BMT) is in 2000 aan de naam van de vereniging toegevoegd om de toenemende belangstelling voor BMT in Nederland tot uitdrukking te brengen. Vanaf 1997 zijn BMT opleidingen gestart aan de Technische Universiteit Eindhoven, de Universiteit Twente en de Rijksuniversiteit Groningen. Per jaar beginnen er meer dan 200 studenten aan een van deze studies. Uniek is de benadering waarbij vanaf het begin van de BMT opleiding de levenswetenschappen worden geïntegreerd in de studie en verschillende componenten van Ontwerp Gericht Leren, het ingenieurs equivalent van Probleem Geörienteerd Leren. BMT is de eerste ingenieurs opleiding tussen de klassieke ingenieurs opleidingen en de levenswetenschappen. De verwachting is dat er meer zullen volgen. Recente ontwikkelingen zijn de opleidingen Medisch Ingenieur (Universiteit Eindhoven) en Technische Geneeskunde (Universiteit Twente).

Onderzoek op het gebied van levenswetenschappen krijgt een steeds prominentere plaats en zij wordt tegelijkertijd steeds afhankelijker van bijdragen vanuit de biomedische technologie en de biofysica. In Nederland is, net als in de hele wereld, een verschuiving zichtbaar richting genetisch en cellulair onderzoek. Biomedische technologie en biofysica zullen cruciaal zijn bij het oplossen van de bijkomende problemen. Deze expertise zal ook voor het begrijpen van organen en van levende wezens van wezenlijk belang blijven, net als voor het begrijpen en ontwikkelen van meetmethoden, zoals ze dat natuurlijk al decennia lang is. Het is dan ook van het grootste belang dat nationale en internationale subsidiegevers biomedische technologie en biofysica als belangrijk gebied erkennen en experts uit deze vakgebieden opnemen in de beoordelingscommissies. Wetenschappers uit Nederland hebben zich altijd gerealiseerd dat internationale samenwerking noodzakelijk is om een lange termijn visie te realiseren waarin fundamenteel onderzoek en onderzoek naar meetmethoden zelf een prominente plaats houden. Nederlandse wetenschapper hebben dan ook een rol gespeeld bij de oprichting en het besturen van verschillende internationale professionele organisaties zoals de IFMBE en de EAMBES, zowel als lid van het dagelijks bestuur als van de raden en werkgroepen.

Nederland heeft een traditie van intensieve samenwerking tussen artsen enerzijds en fysici en ingenieurs anderzijds. Dit heeft geleid tot een goed begrip van de onderliggende fysische principes en van de (patho)fysiologische processen alsmede tot de ontwikkeling van apparatuur om de relevante gegevens te meten. De huidige trend is dat aan de ene kant expertise en traditionele ontwerp-kennis en vakmanschap aan het verdwijnen zijn, maar dat aan de andere kant nieuwe ontwerptechnieken en 'rapid prototyping' technieken opkomen die minder afhankelijk zijn van vakmanschap maar meer van software en computers. Welke gereedschappen we ook in de toekomst zullen gebruiken, de kennis en ervaring van biomedische technologie en biofysica zal altijd een cruciale rol spelen in het onderzoek binnen de levenswetenschappen.

A.C. Linnenbank en D.W. Slaaf

The Nigerian Institute for Biomedical Engineering in (NIBE)

Biomedical Engineering (BME) practice in Nigeria can be said to have started in the 1970s, during which collaborative efforts were made by engineers, medical doctors, pharmacists, physicists, technicians, and other scientists. However, its development has been slow with most of the developmental efforts in the area of training - short courses, continuing education, or professional development. The coming of NIBE, therefore, has propelled the steady progress of BME development in Nigeria, because Nigerian BME professionals started addressing issues concerning them as a professional group. This was given a big boost in 2007, when the first undergraduate programme in BME started in the Federal University of Technology, Owerri, Nigeria, with NIBE contributing the founding members of the faculty.

NIBE, the Nigerian Institute for Biomedical Engineering, represents the BME profession and its members in Nigeria and in international organisations. Established in 1999, with the vision "to develop and advance the biomedical science, health and human well-being of Nigeria through modern technological approaches comparable to those obtainable in any developed country of the world", NIBE has members largely of the various sciences and classical engineering disciplines from universities and hospitals as well as from other institutions/organisations. It has a membership list of over 400 members, but financially active membership fluctuates between 50 and 100 depending on the activity of each year.

NIBE is currently structured in 4 divisions - biological engineering; medical engineering; clinical engineering; and biomedical physics and allied sciences - to accommodate virtually every field in the sciences. It had its 1st annual BME conference in 2000. Since then, it has organised 9 national biomedical engineering conferences and 6 national professional development courses in Nigeria. NIBE has published a newsletter, *news@nibe* since 2000; a professional journal, *Nigerian Journal of Biomedical Engineering*, since 2001; and electronic mail news, *e-nibe*, since 2003.

In 2003, NIBE was admitted as the 50th member of IFMBE. The same year, NIBE co-founded the African Union of Biomedical Engineering and Sciences (AUBES) in Ghana while some members were on a Medical Equipment Training. AUBES was established in order to integrate the effort of various African BME professionals and to expand cooperation on a continental basis. Since 2003, NIBE has made effort with AUBES to pioneer the development of BME in Africa. The official inauguration and the 1st African Biomedical Conference, scheduled for 2005 in Nigeria, was cancelled for logistical reasons. However, a sustained effort is being geared towards its official inauguration in 2010, in Nigeria.

The role of NIBE in developing BME is mainly as a membership group to develop resources for BME by evolving adequate training programmes for members, facilitating accreditation, and certification of professionals practicing BME in Nigeria. NIBE is expected to give its members a sense of belonging by becoming the mouth piece of BME professionals in Nigeria. In the area of education, NIBE is to coordinate the regulation of BME education in Nigeria. All these development efforts should be a matter of concern for the international BME community considering Nigeria's leadership role in Africa. In line with this, NIBE direly needs the support and encouragement of IFMBE and IUPESM as well as WHO and other international bodies involved in BME.

Polish Organizations for Biomedical Engineering Affiliated to IFMBE

There are two national organizations in Poland working in the area of biomedical engineering and affiliated to IFMBE

1. Polish Scientific and Technical Committee for Biomedical Engineering of SEP
2. Polish Society for Biomedical Engineering

The older one is the first organization based on specific expertise. The second organization represents a broader membership including a broader range of professions – technical, medical, and biological.<

Polish Scientific and Technical Committee for Biomedical Engineering of SEP

PKIB

00-050 Warszawa, ul. Świętokrzyska 14
Tel.(+22) 5564-302, 304, Fax: (+22) 5564-301
e-mail: sep@sep.com.pl WWW: www.sep.com.pl

The Polish Scientific and Technical Committee for Biomedical Engineering of Polish Association of Electrical Engineers (PSTCBME of SEP) was established in 1977, (before there was the Medical Electronics Group in the Electronics Section). The Committee was affiliated to IFMBE during the World Congress of MP&BME in Helsinki in 1985. The Committee was directed by prof. Juliusz Keller through 1994, and, since 1994, by prof. Tadeusz Pałko, the president and general secretary of the Committee. In the period 1988 to 1994, prof. Maciej Nalecz, representing the Committee, was a member of the Administrative Council of IFMBE.

The Committee consists of a number of highly qualified experts. The activity of the Committee was concentrated in a wide range of areas of Biomedical Engineering such as: ultrasound technology, intensive care instrumentation, heomodynamic parameter measurement, imaging technology, safety problems, and others. Every year, the Committee organizes three or four meetings and one National Conference of Biomedical Engineering devoted to chosen problems such as non-invasive methods for medical diagnosis and therapy, interventional medicine, anatomical, and functional imaging of human organs.

The Committee cooperates with the International Centre of Biocybernetics (director – prof. M. Nalecz), the International Centre of Hearing and Speech (director – prof. H. Skarzynski), and with the Centre of Excellence on Management of Health and Environmental Hazards (director – dr M. Borysiewicz). The collaboration has resulted in several conferences, seminars, symposia and international workshops.

Polish Society for Biomedical Engineering

02-109 Warsaw, Poland, Ks.Trojdena 4,
Tel.(+22) 658-28-84, Fax: (+22) 659-70-30
e-mail: info@ptib.ibib.waw.pl WWW: http://ptib.ibib.waw.pl

The Polish Society for Biomedical Engineering (PSBE) has been established in November 1998. At present the PSBE has 307 members representing all regions of Poland and various professions, mainly scientists, engineers and physicians. There are also 4 honorary members and 3 foreign members of the society.

The main objectives of the PSBE activity are:

- initiation and promotion of the collaboration between scientists, engineers and physicians,
- organization of conferences, seminars and workshops for education of physicians, biologists and engineers,
- collaboration between national and international societies and institutions,
- supporting the participation of the members in the international projects and research meetings,
- supporting of the editorial activity in the field of biomedical engineering.

To assign its mission the PSBE, besides current activity, is organizing every two years the National Conference on Biocybernetics and Biomedical Engineering and is coediting the quarterly journal *Acta Bio-Optica et Informatica Medica (Inżynieria Biomedyczna)* containing the PSBE News Letters.

Since 2005 the PSBE is affiliated to the IFMBE and the European Alliance for Medical and Biological Engineering (EAMBES).

At present the PSBE Council members are:

Prof. Roman Maniewski (president), Prof. Walerian Staszkiewicz (v-ce President), Assos. Prof. Ewa Zalewska (Secretary General), Dr. Maciej Kozarski (Treasurer) and Prof. Halina Podbielska, Prof. Antoni Nowakowski, Dr. Adam Gacek.

Stowarzyszenie Elektryków Polskich Polski Komitet Inżynierii Biomedycznej

PKIB

00-050 Warszawa, ul. Świętokrzyska 14
Tel.(+22) 5564-302, 304, Fax: (+22) 5564-301
e-mail: sep@sep.com.pl **WWW**: www.sep.com.pl

Komitet Inżynierii Biomedycznej przy Stowarzyszeniu Elektryków Polskich (SEP) został powołany w 1977 roku (wcześniej działał Zespół Elektroniki Medycznej w Sekcji Elektroniki oraz Komitet Inżynierii Biomedycznej NOT). Komitet był afiliowany w 1985 do IFMBE (International Federation for Medical and Biological Engineering) w czasie Światowego Kongresu Fizyki i Inżynierii Biomedycznej w Helsinkach. Komitet był od początku istnienia do 1994 roku kierowany przez prof. Juliusza Kellera a od lutego 1994 do chwili obecnej przez prof. Tadeusza Pałko, prof. M. Nałęcz w latach 1988-1994 jako reprezentant Komitetu był członkiem Administrative Council w IFMBE.

Komitet składa się z ekspertów z obszaru inżynierii biomedycznej. Aktywność Komitetu koncentruje się na szerokim zakresie problematyki inżynierii biomedycznej jak techniki ultrasonograficzne, urządzenia do intensywnej opieki medycznej, pomiary parametrów hemodynamicznych, techniki obrazowania, problemy bezpieczeństwa i inne. Każdego roku organizowane są 3 lub 4 zebrania oraz jedna Konferencja krajowa poświęcona wybranym zagadnieniom np. metodom nieinwazyjnym lub małoinwazyjnym dla diagnostyki i terapii w medycynie interwencyjnej, technikom obrazowania funkcjonalnego i anatomicznego. Komitet współpracuje z Instytutem Biocybernetyki i Inżynierii Biomedycznej w tym z Międzynarodowym Centrum Biocybernetyki (dyrektor – prof. M. Nałęcz), Międzynarodowym Centrum Słuchu i Mowy (dyrektor - prof. H. Skarżyński), Centre of Excellence on Management of Health and Environmental Hazards (dyrektor – dr M. Borysiewicz).

Polskie Towarzystwo Inżynierii Biomedycznej

02-109 Warszawa, Ks.Trojdena 4,
Tel.(+22) 658-28-84, Fax: (+22) 659-70-30
e-mail: info@ptib.ibib.waw.pl WWW: http://ptib.ibib.waw.pl

Polskie Towarzystwo Inżynierii Biomedycznej (PTIB) zostało utworzone w listopadzie 1998 roku. Aktualnie PTIB ma 307 członków reprezentujących wszystkie regiony Polski, zarówno środowiska pracowników naukowych jak i placówki medyczne i instytucje związane z przemysłem aparatury medycznej. Towarzystwo ma także 4 członków honorowych oraz 3 członków zagranicznych.

Zgodnie ze Statutem PTIB głównymi celami działalności Towarzystwa są:

- inicjowanie i promowanie współpracy pomiędzy przedstawicielami nauk technicznych, medycznych i przemysłem,
- organizacja posiedzeń naukowych i dyskusyjnych, odczytów, wykładów, zjazdów naukowych i dydaktycznych, szkół letnich i innych form szkolenia dla lekarzy, biologów i inżynierów,
- współpraca z krajowymi i zagranicznymi instytucjami i towarzystwami naukowymi,
- wspieranie uczestnictwa w krajowych i międzynarodowych konferencjach,
- inspirowanie i wspieranie publikacji naukowych i działalności wydawniczej,

Dla osiągnięcia powyższych celów PTIB, oprócz bieżącej działalności, organizuje co 2 lata Krajową Konferencję Naukową *Biocybernetyka i Inżynieria Biomedyczna.* Od początku działalności PTIB wydaje Biuletyn Informacyjny PTIB, który jest dołączany do wydawanego przy współudziale PTIB kwartalnika *Acta Bio-Optica et Informatica Medica (Inżynieria Biomedyczna).*

Od roku 2005 PTIB jest członkiem międzyanarodowych federacji towatrzystw naukowych IFMBE oraz European Alliance for Medical and Biological Engineering (EAMBES).

Aktualny Zarząd Główny został wybrany w roku 2007 w następującym składzie:

Prof. Roman Maniewski (Prezes), Prof. Walerian Staszkiewicz (v-ce Prezes), Doc. Ewa Zalewska (Sekretarz), Dr. Maciej Kozarski (Skarbnik) oraz członkowie: Prof. Halina Podbielska, Prof. Antoni Nowakowski, Dr. Adam Gacek.

Portuguese Society of Biomedical Engineering (SPEB)

The Portuguese Society of Biomedical Engineering (SPEB) was created in January, 25th, 1989, almost one year after the first Portuguese Biomedical Engineering conference, BioEng'88, took place. At the time of creation the society had 89 members and was essentially an association of people doing research in Biomedical Engineering, with special emphasis in Biomedical signal processing and Biomaterials. Since then eight national Biomedical Engineering conferences took place and the society has had four different directive boards, with the present one incorporating people from most major universities in the country with Biomedical engineering groups. The areas of interest have grown, including, among others, Biomechanics, Medical Imaging, Medical Informatics and Tissue Engineering. In 2001, the first undergraduate degrees in Biomedical Engineering were created in Portugal and the Portuguese BME community was faced with a new challenge: the appearance of a new professional career in Biomedical Engineering. At the moment there are 8 universities and 4 polytechnics offering undergraduate Biomedical Engineering degrees.

Portugal still does not have professional Clinical Engineers, and their acceptance should be one of the challenges of SPEB. The Portuguese society has been affiliated to IFMBE for many years and recently has become more actively involved in the Federation.

For the near future SPEB plans to become a more representative society in the field with several actions planned which involve: increasing the number of its members, mainly from the undergraduate and newly graduated students; create a Biomedical Engineering Education and certification group with links to BIOMEDEA; create contacts with the various BME companies in the country; establish a more regular Portuguese BEM conference; strengthen the links with other societies, like the Medical Physics Division of the Portuguese Physical Society and the Portuguese Biomechanics Society; fight for the creation of a Clinical Engineering profession in Portugal and its acceptance within the medical community; strengthen even more the links with European and International organisations.

Sociedade Portuguesa de Engenharia Biomédica (SPEB)

A Sociedade Portuguesa de Engenharia Biomédica (SPEB) foi criada a 25 de Janeiro de 1989, quase um ano depois da primeira Conferência Portuguesa de Engenharia Biomédica, BioEng'88, ter ocorrido. Na altura da sua criação a Sociedade tinha 89 membros e era essencialmente uma associação de pessoas que faziam investigação em Engenharia Biomédica, com especial ênfase em Processamento de Sinal Biomédico e Biomateriais. Desde então já tiveram lugar oito conferências nacionais de Engenharia Biomédica e a sociedade teve quatro Direcções diferentes, incorporando a Direcção actual pessoas da maioria das universidade do país com grupos de Engenharia Biomédica. As áreas de interesse cresceram, incluíndo, entre outras, Biomecânica, Imagem Médica, Informaática Médica e Engenharia de Tecidos. Em 2001 os primeiros cursos de primeiro ciclo de Engenharia Biomédica foram criados em Portugal e a comunidade de EBM deparou-se com um novo desafio: o aparecimento de uma nova carreira profissional em Engenharia Biomédica. De momento há 8 universidades e 4 politécnicos a oferecerem cursos de Engenharia Biomédica a nível do primeiro ciclo. Portugal ainda não tem Engenheiros Clínicos profissionais e a sua aceitação deve ser um dos desafios da SPEB.

A sociedade Portuguesa está afiliada à IFMBE há mitos anos e recentemente tornou-se mais activamente envolvida na Federação.

Para um futuro próximo a SPEB planeia tornar-se uma sociedade mais representativa nesta área com várias acções planeadas que envolvem: aumentar o número dos seus membros, principalmente com alunos a estudar e recentemente formados; criar um grupo de educação e certificação em EBM com ligações à BIOMEDEA; criar contactos com as várias empresas de EBM no país; estabelecer uma conferência portuguesa de EBM mais regular; fortalecer as ligações com outras sociedades, como a Divisão de Física Médica da Sociedade Portuguesa de Física e a Sociedade Portuguesa de Biomecânica; lutar pela criação da profissão de Engenheiro Clínico em Portugal e pela sua aceitação na comunidade médica; fortalecer ainda mais as ligações com as organizações europeias e internacionais.

Romanian National Society of Medical Engineering and Biological Technology (NSMEBT)

The **Romanian National Society of Medical Engineering and Biological Technology (NSMEBT)** was established on June 2nd, 2000 when 49 founders were present and was legally recorded on March 30th, 2001.

The NSMEBT is an independent scientific organisation, with an interdisciplinary background, aiming at promoting the co-operation, research, application of know-how and dissemination of information related to the utilisation of modern technology in the field of medical, clinical and biological engineering.

The serious commitment of our Society to education is also shown by the fact that the majority of its members are university staff; other members work in research institutes, hospitals, clinics, and health care departments, private companies trading medical equipment or offering service for the equipment.

An important role is played by the Romanian Society of Medical Engineering and Biological Technology both in the promotion of biomedical engineering in Romania and in the education and training of the individuals interested in biomedical engineering and science.

Teaching activities started in Romania, at the university level, in 1997, when a new Department of Medical Techniques was established at the Electrotechnical Faculty of the Technical University of Cluj-Napoca. Since then similar educational programmes have been established at other Romanian technical universities.

Project-based learning is a way to ensure that the students not only have basic knowledge, but also learn to work with that knowledge, to apply it, basically developing this skill to solve problems, and to do so within a group.

The NSMEBT has organised different activities (scientific mainly); some examples: the international conference "Advancements of Medicine and Health Care through Technology – MediTech" 2007 (the 1st edition), 2009 (the 2nd edition), the training "The use of ANSYS-CFX software in medical applications" (March 2008; trainer; Dr. Andrew Narracot, UK), etc.

The Executive Council of NSMEBT decided in 2009 to reduce the membership fee for 2010 for applicants younger than 35 by 50% in order to stimulate their involvement in the activities of the National Society.

In January 2009 the Romanian National Society of Medical Engineering and Biological Technology applied for IFMBE membership, which has been accepted.

The partnership with the IFMBE will enhance the research and educational activities in the field of biomedical engineering in Romania. Furthermore, this partnership will create opportunities for collaboration with national societies which are members of IFMBE.

Societatea Nationala de Inginerie Medicala si Tehnologie Biologica (SNIMTB)

Societatea Nationala de Inginerie Medicala si Tehnologie Biologica (SNIMTB) din Romania a luat fiinta in anul 2000 prin decizia a 49 de membri fondatori si s-a constituit legal prin decizie judecatoreasca in 30 martie 2001.

SNIMTB este o organizatie stiintifica independenta care urmareste promovarea cooperarii in cercetare, sprijinirea invatamantului de profil si diseminarea informatiilor in ceea ce priveste utilizarea tehnologiei moderne in domeniile ingineriei medicale, clinice si biologice.

Implicarea serioasa a societatii noastre in educatie este demonstrata si de faptul ca majoritatea membrilor ei sunt cadre didactice universitare; alti membri lucreaza in institute de cercetare, spitale, clinici, departamente de ingrijirea sanatatii, companii private angajate in producerea sau service-ul de echipamente medicale.

SNIMTB joaca un rol important pe de o parte in promovarea ingineriei biomedicale in Romania, iar pe de alta parte in educarea si pregatirea celor implicati sau preocupati de stiinta si ingineria biomedicala.

Educatia in acest domeniu, la nivel universitar a inceput in 1997 cand s-a infiintat specializarea de Tehnica Medicala (colegiu) in cadrul Facultatii de Electrotehnica a Universitatii Tehnice din Cluj-Napoca. De atunci programe educationale similare au luat fiinta si in alte Universitati Tehnice din tara.

SNIMTB a organizat o serie de activitati (in principal stiintifice) dintre care amintim: Conferinta Internationala "Advancements of Medicine and Health Care through Technology – MediTech", 2007 – editia I-a, 2009 – editia a II-a, Workshop "The use of ANSYS-CFX software in medical applications" – martie 2008 (Dr. Andrew Narracot, UK).

Colegiul Director al societatii a decis in 2009 reducerea cuantumului cotizatiei de membru pentru anul 2010 cu 50% pentru cei sub 35 de ani in scopul stimularii implicarii lor in activitatile societatii.

In ianuarie 2009 SNIMTB si-a depus candidatura de membru al IFMBE, care a fost acceptata. Colaborarea cu IFMBE va permite revigorarea activitatii de cercetare si educationale in domeniul ingineriei medicale in Romania, ca si crearea de oportunitati de colaborare cu societatile membre ale IFMBE.

The Society of Biomedical Engineering and Medical Physics of Serbia

The Society of Biomedical Engineering and Medical Physics of Serbia (BIMEF) was founded in late 1980s as a voluntary scientific - professional national organization consisting of qualified biomedical engineers and medical physicists with university degrees and working in alliance with medical staff in hospitals, universities, or research institutes. BIMEF is a member of European federation of medical physicists (EFOMP) and IFMBE. The aims and tasks of the BIMEF include:

- Encouraging and promulgating scientific work and technology development in the fields of activity of the members
- Follow the development and application of biomedical engineering and medical physics methods in all country centers
- Participation in education, training, and in CPD accreditation programs
- Disseminating professional and scientific information through publications and meetings
- Provide professional consultation in the field of activity
- Publishing professional scientific publications
- Organization and participation in work of national or international conferences in the field of activity
- Collaboration with national and international organizations with the same field of interest
- Additional education of engineers, physicists, and technicians in the field of biomedical engineering and medical physics
- Organizing courses, seminars, and other kinds of meetings.

After the long break, in 2004, BIMEF restarted its activities. The most important effort was the International workshop dedicated to Medical Physics in South-Eastern Europe. The main aim of this conference was to affirm the place of medical physicists and clinical engineers in hospitals, to find a way for these professions to be recognized as registered occupations, and to assign responsibilities and establish means of licensing and the permanent education of medical physicists. Additionally, at the same time, the First International School dedicated to Treatment Planning in Radiotherapy was organized. There were 64 participants and lecturers from Serbia and Montenegro, FYR Macedonia, Cyprus, Hungary, Austria, Romania, Italy, Germany, and Switzerland.

The Second International School was organized in 2005, with a 4 day programme dedicated to Radiation protection in radiotherapy, radiological diagnostics, and nuclear medicine. There were 48 participants and lecturers from Serbia and Montenegro, FYR Macedonia, Cyprus, Austria, Slovenia, Bosnia, and Hercegovina.

The Third International School, recognized as a joint meeting of the Serbian and FYR Macedonian societies, was sponsored by EFOMP and organized in November 2008. The main topics of the school covered advantageous techniques in radiological diagnostics. The 2 day programme of Quality Assurance and Quality Control in Diagnostic Imaging was proposed by EFOMP Officers who also participated as lecturers. The school had 74 lecturers and participants from Italy, Serbia, Montenegro, FYR Macedonia, Bosnia, Hercegovina, and Austria. Besides the professionals from hospitals, research institutes, private clinics, and universities, we also had a significant number of our students from University of Novi Sad, from the Faculty of Technical Sciences and Faculty of Science.

In the future, we shall continue to put all our efforts to promote our stated tasks and goals. Our future tasks will be pointed toward the recognition of the profession as well as to the accreditation of all necessary CPD programmes. The main problem is to find how to put the professions of medical engineer and medical physicist on the Registered list of profession. The solution to this problem will be of vital importance for the profession and our future work. We shall continue with the organization of international schools, workshops, and meetings. We also plan to join with the Faculty of Technical Sciences in TEMPUS IV Second application, with the aim to reform the existing syllabus and the place of higher education in society. In the next 4 years, we also plan to publish a number of publications as a training and teaching material for our members.

Prof.dr. Vesna Spasić-Jokić,
President,
Faculty of Technical Sciences,
University of Novi Sad

Društvo za biomedicinsko inženjerstvo i medicinsku fiziku Srbije

Društvo za biomedicinsko inženjerstvo i medicinsku fiziku Srbije, BIMEF, je osnovana osamdesetih godina kao dobrovoljna naučno-stručna organizacija kvalifikovanih biomedicinskih inženjera i medicinskih fizičara sa univerzitetskim diplomama, koji su radili u zdravstvu, na univerzitetima i u naučnim institutima. BIMEF je član evropske federacije EFOMP i međunarodne federacije IFMBE. Ciljevi i zadaci BIMEF uključuju: podsticanje i širenje naučnog rada i tehnološkog razvoja u oblastima za koje je registrovan; praćenje razvoja i aplikacije metoda u oblasti biomedicine i medicinske fizike u nacionalnim centrima; učešće u edukaciji, obuci i akreditacionim programima za kontinualni profesionalni razvoj CPD; prenošenje profesionalnih i naučnih informacija kroz publikacije i sastanke; pružanje stručnih konsultacija u oblastima relevantnim za aktivnost članstva; publikovanje svih vrsta naučnih publikacija vezanih za delatnost društva; organizacija i učešće u organizaciji nacionalnih i međunarodnih konferencija, radionica i seminara; saradnja sa nacionalnim i međunarodnim srodnim društvima i organizacijama; dopunska edukacija inženjera, fizičara i tehničara u oblasti biomedicinskog inženjerstva i medicinske fizike kao i organizacija kurseva, seminara i tribina.

Posle duže pauze BIMEF je 2004. godine nastavio sa svojim aktivnostima. U okviru planiranih aktivnosti za 2004. najznačajnija je bila međunarodna radionica *Medical Physics in South-Eastern Europe*. Osnovni cilj ovog skupa je bio da afirmiše mesto medicinskih fizičara i kliničkih inženjera u bolnici i da predloži načine kako da se ove profesije registruju, kako da se ustanove odgovornosti, načini licenciranja i programi permanentne edukacije. Kao dodatak ovoj radionici organizovana je Prva međunarodna škola posvećena planiranju u radioterapiji, *Treatment Planning in Radiotherapy*. Na oba skupa održana u dva dana bilo je 64 predavača i učesnika iz Srbije i Crne Gore, Kipra, Italije, Austrije, Mađarske, BJR Makedonije, Rumunije, Nemačke i Švajcarske.

Druga međunarodna škola sa četvorodnevnim programom je organizovana 2005. godine i bila je posvećena zaštiti od zračenja u medicini, *Radiation protection in radiotherapy, radiological diagnostics and nuclear medicine*. Bilo je 48 učesnika i predavača iz Srbije i Crne Gore, Slovenije, Kipra, Austrije, Bosne i Hercegovine i BJR Makedonije.

Treća međunarodna škola je organizovana kao zajednički regionalni sastanak Srpskog i BJR Makedonskog društva. Škola je bila sponzorisana od strane EFOMP i održana je 2008. Glavne teme škole su pokrivale neke napredne tehnike u radiološkoj dijagnostici. Naslov je bio *Quality Assurance and Quality Control in Diagnostic Imaging*. Dvodnevni ambiciozni program je bio predložen od strane funkcionera EFOMP, koji su i učestvovali kao predavači. Škola je imala 74 predavača i učesnika iz Italije, Srbije, Austrije, Bosne i Hercegovine, Crne Gore, BJR Makedonije Među učesnicima se pored stručnjaka iz bolnica, instituta, privatnih klinika i univerziteta našao i značajan broj naših studenata sa Fakulteta tehničkih nauka i Prirodno-matematičkog fakulteta, oba sa Univerziteta u Novom Sadu.

U budućem radu ćemo uložiti napore na daljoj promociji osnovnih ciljeva i zadataka. Naši glavni zadaci će biti fokusirani na prepoznavanje profesije kao i na akreditaciju CPD programa. Glavni problem je da se iznađu načini kako da se profesije klinički inženjer i medicinski fizičar nađu na listi Registra profesija. Rešenje ovog problema je od vitalnog značaja za profesiju i naš budući rad. Nastavićemo sa organizacijom međunarodnih škola i sa izdavačkom delatnošću. Planiramo da se Fakultetu tehničkih nauka pridružimo u aplikaciji za TEMPUS IV u cilju reforme silabusa i mesta visokog školstva u društvu u ovoj oblasti.

Prof.dr. Vesna Spasić-Jokić,
Predsednik
Fakultet tehničkih nauka,
Univerzitet u Novom Sadu

The Biomedical Engineering Society (Singapore)

The idea of forming a biomedical engineering society in Singapore was promoted way back in 1985. It started by a group of enthusiastic engineers and medical doctors from university, polytechnics, industry, and hospitals. However, its actual formation was only finalised in 1991, when approval was granted by our Registry of Societies for the formation of the Biomedical Engineering Society (Singapore) or BES. The Society started with about 30 members in 1991, and has grown to about 65 regular members and about 40 student members. The Society has recently introduced corporate membership to attract participation of the biomedical engineering companies in Singapore.

The Society has achieved most of the objectives laid out in our constitution.

- In November 2001, our Society's views were sought by the Bioethics Advisory Committee for feedback regarding human stem cell research in Singapore.
- In November 2002, we signed a Memorandum of Understanding (MoU) between the Institution of Engineers, Singapore (IES) and Biomedical Engineering Society (Singapore) (BES). The MoU focused on promoting co-operation and co-ordination of activities and communication between the two organisations, such as conferences, seminars, workshops, and continuing education and programmes.
- To effect scientific exchanges among its members and others, BES has been involved in organizing the International Conference on Biomedical Engineering (ICBME) since 1997, when it co-organised the 9th ICBME with the National University of Singapore. The 13th ICBME was held on 3-6 December 2008, in Singapore, and the next one will be our 14th ICBME from 1-6 August 2010. This conference will be held under the umbrella of the 6th World Congress on Biomechanics which is also organized by our Society.
- Together with the Thai Biomedical Engineering Society, BES also co-organised the WACBE World Congress on Bioengineering on 9 – 11 July 2007.
- In addition, the Society has also organized 4 scientific meetings for our members targeted mainly to our student members as well as non-members.
- We have organized technical seminars whenever we have prominent people in the field of biomedical engineering in town like Prof. Shu Chien of University of California San Diego, Prof. Robert Nerem of Institute of Bioengineering and Biosciences, Georgia Tech, Georgia, USA and Prof. Edmund Chao from the Orthopaedic Biomechanics Lab, Johns Hopkins University, just to name a few.

The Slovak Society of Biomedical Engineering and Medical Informatics

The Slovak Society of Biomedical Engineering was established in 1969, as one of the professional societies belonging to the Slovak Medical Association. It was founded by engineers, biophysicists, and medical doctors from universities, research and development institutions, and from clinical practice. In 1996, it was transformed into the Slovak Society of Biomedical Engineering and Medical Informatics, and two working sections, Biomedical Engineering (BME) and Medical Informatics (MI), were established. Since its origin, the Society has been active mainly in the scientific and educational fields, but, in recent years, an expert and legislative activities that became more important in connection with the health care system transformation in the country.

In 2008, the Society had over 60 active members, and more than 30 BME specialists and students who cooperated with the Society on an informal basis. Currently, most of the members are from universities and research institutions, some of whom are clinical engineers and practising medical doctors. The Society is the official representative of the Slovak biomedical engineering and medical informatics community in the International Federation for Medical and Biological Engineering (IFMBE) and International Medical Informatics Association (IMIA).

Activities of the Society and its members are closely related to research and education in BME in the former Czechoslovakia and later in Slovakia. In the late 1960s, research in BME began in the Institute of Measurement Science of the Slovak Academy of Sciences in Bratislava and the development of medical electronics was established in the medical electronics company Chirana in Stara Tura. In 1971, an educational programme for medical electronics was established at the Faculty of Electrical Engineering, Slovak University of Technology in Bratislava. In 1987, biomedical engineering study oriented mainly to biomechanics and rehabilitation engineering was initiated at the Faculty of Mechanical Engineering, Technical University of Košice. Since 1999, an interdisciplinary study of Biomedical Engineering was prepared as a result of a Tempus/Phare programme at the Faculty of Electrical Engineering, University of Zilina, in cooperation with Jesenius Medical Faculty, Comenius University in Martin. In 2004, the Centre of Biomedical Engineering, oriented mainly to education and research in biosensors and biomaterials, was established at the Slovak University of Technology in Bratislava. Besides the undergraduate and graduate programmes, postgraduate courses have been organised since the 1980s, in cooperation between the universities and the Slovak Academy of Sciences.

One of the main activities of the Society is organisation of national and international conferences, seminars, and workshops in biomedical engineering and medical informatics. Let us mention just few of them. In close cooperation with universities and with the Czech Society for Biomedical Engineering and Medical Informatics, seven conferences "Trends in biomedical engineering" and three biannual "YBERC" (Young Biomedical Engineers and Researchers) conferences were organised. In cooperation with the Institute of Measurement Science SAS, sessions devoted to measurements in biomedicine were prepared at six international "Measurement" conferences.

For many years, publishing activities of the Society (as a co-publisher) were concentrated around the journal *Lekar a technika* (*Physician and Medical Technology*) published in Czech and Slovak languages by the Czech Society as the main publisher.

The Society has its web page at http://www.sbimi.sk that brings information on events in the BME and MI community, mediates participation of its members in expert and legislative activities, and informs about the Society actions.

Besides these traditional activities, the future goals of the Society are for stronger stimulation of publications of students and young engineers, extension of the use of the web page as a discussion forum to actual problems of the community, and involvement in preparation and organisation of e-learning courses in BME and MI in Slovakia.

Biosignal Conference 2006 Opening

Slovenská spoločnosť biomedicínskeho inžinierstva a medicínskej informatiky

Slovenská spoločnosť biomedicínskeho inžinierstva vznikla v roku 1969 ako jedna z odborných spoločností patriacich do Slovenskej lekárskej spoločnosti. Bola vytvorená inžiniermi, biofyzikmi a lekármi z univerzít, výskumných a vývojových ústavov a z klinickej praxe. V roku 1996 sa transformovala na Slovenskú spoločnosť biomedicínskeho inžinierstva a medicínskej informatiky a boli vytvorené dve sekcie: sekcia biomedicínskeho inžinierstva (BMI) a sekcia medicínskej informatiky (MI). Od vzniku spoločnosti sa jej činnosť zameriava najmä na vedecké a vzdelávacie aktivity, ale v posledných rokoch nadobúdajú význam aj ďalšie expertné a legislatívne aktivity súvisiace s transformáciou systému zdravotnej starostlivosti na Slovensku.

V roku 2008 mala spoločnosť vyše 60 aktívnych členov a viac ako 30 špecialistov a študentov z oblasti BMI a MI so spoločnosťou spolupracovalo. V súčasnosti je väčšina členov spoločnosti z univerzít a výskumných pracovísk, ďalší pracujú ako klinickí inžinieri alebo lekári. Spoločnosť je oficiálnym predstaviteľom slovenskej komunity biomedicínskych inžinierov a medicínskych informatikov v medzinárodných spoločnostiach International Federation for Medical and Biological Engineering (IFMBE) a International Medical Informatics Association (IMIA).

Aktivity spoločnosti a jej členov úzko súvisia s výskumom a vzdelávaním v oblasti BMI v bývalom Československu a neskôr v Slovenskej republike. Výskum v oblasti BMI sa začal koncom šesťdesiatych rokov minulého storočia v Ústave merania Slovenskej akadémie vied v Bratislave. V tomto čase tiež vzniklo vývojové oddelenie lekárskej elektroniky v podniku na výrobu zdravotníckej techniky Chirana Stará Turá. Od roku 1971 sa začalo na Elektrotechnickej fakulte Slovenskej technickej univerzity v Bratislave štúdium so zameraním na lekársku elektroniku. V roku 1987 bolo zahájené štúdium biomedicínskeho inžinierstva orientované najmä na biomechaniku a rehabilitačné inžinierstvo na Strojníckej fakulte Technickej univerzity v Košiciach. Od roku 1999 sa rozbehlo interdisciplinárne štúdium biomedicínskeho inžinierstva, ktoré bolo pripravené ako výsledok projektu Tempus/Phare na Elektrotechnickej fakulte Žilinskej univerzity v spolupráci s Jeseniovou lekárskou fakultou Univerzity Komenského v Martine. A nakoniec v roku 2004 bolo založené Centrum biomedicínskeho inžinierstva orientované najmä na vzdelávanie a výskum v oblasti biosenzorov a biomateriálov na Slovenskej technickej univerzite v Bratislave. Okrem bakalárskych a inžinierskych študijných programov sa od osemdesiatych rokov minulého storočia v spolupráci univerzít a Slovenskej akadémie vied rozbehlo aj doktorandské štúdium.

Jednou z hlavných aktivít spoločnosti je organizácia národných a medzinárodných konferencií, seminárov a pracovných stretnutí v oblasti biomedicínskeho inžinierstva a medicínskej informatiky. Spomeňme aspoň niektoré z nich. Spoločnosť v úzkej spolupráci s univerzitami a s Českou spoločnosťou biomedicínskeho inžinierstva a lekárskej informatiky zorganizovala sedem ročníkov konferencie "Trendy v biomedicínskom inžinierstve" a tri dvojročné konferencie "YBERC" (Young Biomedical Engineers and Researchers Conference). V spolupráci s Ústavom merania SAV boli pripravené sekcie venované meraniu v biológii a medicíne na šiestich medzinárodných konferenciách "Measurement".

Publikačné aktivity spoločnosti (ako spoluvydavateľa) boli po mnoho rokov sústredené okolo časopisu Lekár a technika, ktorý vychádzal v češtine a slovenčine a ktorého hlavným vydavateľom bola Česká spoločnosť biomedicínskeho inžinierstva a lekárskej informatiky.

Spoločnosť má vlastnú webovú stránku http://www.sbimi.sk, ktorá informuje o udalostiach v oblasti BMI a MI, sprostredkúva členom spoločnosti účasť v expertných a legislatívnych činnostiach a informuje o aktivitách spoločnosti.

Okrem týchto tradičných aktivít, medzi budúce zámery spoločnosti patrí silnejšia stimulácia publikačnej činnosti študentov a mladých inžinierov, rozšírenie využitia webovej stránky ako diskusného fóra k aktuálnym problémom a zaangažovanie sa do príprav a organizácie e-learningových kurzov v oblasti BMI a MI.

The Slovenian Society for Medical and Biological Engineering

The Slovenian Society for Medical and Biological Engineering was founded in 1986. The inaugural meeting of the Society was held in Ljubljana, February 2, 1986.

At that time, the Slovenian Society was established as a member society of the federal Yugoslav society for medical and biological engineering. With Slovenia becoming an independent state, the Society started activities for becoming a national member of the IFMBE. This decision was taken on October 15, 1991, in Portschachu, Austria, where the 7th International Conference on Mechanics in Medicine and Biology (13 to 16 October 1991) met, due to the war in Slovenia. The Slovenian Society submitted the Membership Application February 5, 1992, and was formally accepted as a member of the IFMBE at the General Assembly during the world congress in Rio de Janiero, 1994.

The Slovenian Society from its very beginning brought together biomedical engineers and different professionals such as medical doctors, engineers, physicists, biologists, and others working in heath care. Also, the Society has been publishing its *Newsletter*, from 1988 until 2001 in paper form, and later in electronic form. The Society is also organizing national and international conferences.

Conferences organized by Slovenian Society for Medical and Biological Engineering:

7th International Conference on Mechanics in Medicine and Biology Ljubljana - Portschach (13 - 16 October 1991); (then still as a member of the Yugoslav society)

The Ljubljana FES Conference, Ljubljana, Slovenia, (August 22-25, 1993)

2nd Congress of The European Bioelectromagnetics Association, Bled-Slovenia (December 9 – 11, 1993)

9th International Conference on Mechanics in Medicine and Biology, Ljubljana, Slovenia (June 30 – July 4, 1996)

3rd International Conference on Bioelectromagnetism and 1st Slovenian – Croatian Meeting on Biomedical Engineering, Bled – Slovenia, (8 – 12 October 2000)

11th Mediterranean Conference on Medical and Biological Engineering and Computing 2007, 26 – 30 June, Ljubljana, Slovenia

In the last two years, members of the Society, who are also professors at the University of Ljubljana, are preparing a self-standing educational programme of Biomedical Engineering at the Faculty of Electrical Engineering, where a Department for Biomedical Engineering was formed for this purpose. The programme of Biomedical Engineering for a Master's degree, according to Bologna declaration, is now to be submitted for national accreditation.

Conference excursion to Bled during MEDICON 2007

Slovensko društvo za medicinsko in biološko tehniko

Slovensko društvo za medicinsko in biološko tehniko je bilo ustanovljeno leta 1986. Na ustanovni skupščini društva dne 20. 2. 1986 potrdi tudi statut društva.

Takratno slovensko društvo je bilo ustanovljeno v okviru zveze društev Jugoslovanskega društva za medicinsko in biološko tehniko. Z osamosvojitvijo Slovenije so stekle aktivnosti za osamosvojitev društva in včlanitev v IFMBE. Sklep o začetku teh aktivnosti je bi sprejet na ad-hoc sestanku članov društva 15. 10. 1991 v Portschachu (Avstrija), kamor je bila zaradi vojne v Sloveniji premeščena konferenca 7th International Conference on Mechanics in Medicine and Biology (13 do 16. Oktober, 1991).

Včlanitev v IFMBE – Membership Application February 5, 1992: formalno je bilo slovensko društvo sprejeto na skupščini IFMBE v Rio de Janeiru avgusta 1994.

Društvo že od samega začetka povezuje strokovnjake biomedicinske tehnike, inženirje, zdravnike in druge zaposlene v zdravstvu. Pri tem se poslužuje izdajanja Novic društva in organizacije konferenc – tako nacionalnih kot tudi mednarodnih. Novice društva so pričele izhajati leta 1988 in so v papirni obliki izhajale do leta 2001. Zatem so prešle v elektronsko obliko.

Konference, ki jih je društvo (so)organiziralo:

7th International Conference on Mechanics in Medicine and Biology Ljubljana - Portschach (13 - 16. Oktorber ,1991); takrat še v okviru jugoslovanskega društva

The Ljubljana FES Conference , Ljubljana Slovenija, (August 22-25, 1993)

2nd Congress of The European Bioelectromagnetics Association, BLED-SLOVENIA (December 9 – 11, 1993)

9th International Conference on Mechanics in Medicine and Biology, LJUBLJANA, SLOVENIA (June 30 – July 4, 1996)

\3rd International Conference on Bioelectromagnetism and 1st Slovenian – Croatian Meeting on Biomedical Engineering , Bled – SLOVENIA, (8 – 12 OKTOBER 2000)

11th Mediterranean Conference on Medical and Biological Engineering and Computing 2007, 26 – 30 June, Ljubljana, Slovenia

V zadnjih dveh letih si člani društva, ki so obenem tudi učitelji na Univerzi v Ljubljani, prizadevajo za ustanovitev samostojne smeri Biomedicinska tehnika na Fakulteti za elektrotehniko. V ta namen je bila na Fakulteti ustanovljena Katedra za biomedicinsko tehniko. Trenutno je pripravljena smer Biomedicinska tehnika za 2. bolonjsko stopnjo.

Presentation of Awards to IFMBE Young Investigator Competition winners, Ljubljana 2007, Slovenia

The Biomedical Engineering Society of South Africa (BESSA)

We salute the IFMBE on reaching its 50th year of service. It is for us also a reason to celebrate because 2008 marked the 30th anniversary of BESSA's establishment.

Biomedical Engineering was a young and vital field in South Africa in the 1970s and the first academic programme in BME was established, significantly, at the University of Cape Town Medical School in late 1972, by Mickey Milner (who is well known in the IFMBE as a past president). Mickey and Terry van der Werff helped build up BME in South Africa and their and other enthusiasts' efforts led to the establishment of BESSA in 1977. Shortly afterwards, Stellenbosch University and Pretoria University, followed by Witwatersrand University, started BME streams, all within existing engineering faculties.

The 1980s were the heyday of Biomedical Engineering in South Africa due to the rapid advances in Medical/Clinical Engineering and the consequent need for professionals to develop, manage, and maintain medical equipment – including radiological and radiotherapeutic. BESSA formed a close relationship with the South African Society for Medical Physics; and the National Accelerator Centre with a clinical wing for neutron and proton therapy was established in Faure, near Cape Town. Close liaison between medical physicists and biomedical engineers led to the design and development of an innovative robotic patient positioning system, using real-time video stereophotogrammetry. This approach has since been emulated worldwide.

Initially, Biomedical Engineering higher degree programmes at South African universities were all at graduate level, but introductory/elective courses in Biomedical Engineering were gradually included as electives in undergraduate engineering programmes at the University of Cape Town and Pretoria.

In terms of professional development, it became a mandated legal requirement for BME graduates to be registered with the SA Health Professions Council (HPC) alongside doctors, dentists, nurses, and medical physicists, if they engaged in patient care related activities. Clinical engineering was also developing apace and the South African Society for Clinical Engineering (SAACE) was established in the early 1990s. Close liaison was maintained between the two societies, but the breakaway had the effect of diminishing BESSA's membership and seriously diluting the influence of BME/Medical Engineering as a single lobby to engage with government and influence the healthcare industry in South Africa.

In 2004, Clinical Engineering was professionally accredited by the Engineering Council of South Africa as a specialist branch of the profession. In 1982, BESSA's Constitution was changed to allow Northern and Southern ChapterS to be established. This reflected the realities of South African geography, since the country has two major population/commercial concentrations with a 1000km distance between them. This allowed the two regional BME groups to hold separate meetings locally and to alternate national conferences between the Northern (Johannesburg-Pretoria-Durban) and Southern (Cape Town) Regions.

Due to major cutbacks in funding for Specialist Academic Hospitals in favour of improving Primary Health Care post democratization of South Africa in 1994, academic BME programmes and staffing levels were curtailed and student enrolment and professional career opportunities decreased for a number of years, leading to a sharp decline in BESSA membership, which now sits at some 22 stalwarts – mainly in academe! The establishment, in 2003, of a special new undergraduate engineering science programme in the Witwatersrand University (Johannesburg) School of Engineering, spearheaded by BESSA member David Rubin, was a cause for celebration. It was the first to effectively offer a pre-clinical engineering/science degree, allowing migration paths into science, engineering or medicine. Some 40% of the first graduates from the programme in 2006 went on to enter Medicine.

The realities of natural attrition of members and the fragmenting of Biomedical Engineering interests into academic, clinical engineering, and hospital engineering has resulted in BME having to take on a more management-orientated role. In response, Mladen Poluta, at U.Cape Town, has established a Regional/African- directed diploma/degree programme in Healthcare Technology Management, which enables a wide range of health professionals to acquire the skills and knowledge needed to more effectively utilise technology in healthcare at all levels, both preventative and curative, but with a strong focus on African realities.

In conclusion, BESSA recalls with gratitude the many contributions by visiting lecturers from IFMBE affiliates overseas, who over the years helped us develop new academic programmes or who gave inspiring keynote addresses at Southern African Regional BESSA/IFMBE Scientific Conferences.

David Boonzaier MBChB (UCT), DIC (London)
Secretary/Treasurer

Biomedical Engineering Society of South Africa.

OFFICE BEARERS

Rotating Chairs: Peter Heimann & Mladen Poluta

Secretary/ Treasurer: David Boonzaier

Administrator: Cathy Hole

Spanish Society for Biomedical Engineering (SEIB)

The Spanish Society SEIB was founded in 1981, and it is the Biomedical Engineering Spanish association whose activities extend throughout Spain.

The objectives of SEIB are to:

- Serve and promote Biomedical Engineering (BME) in Spain
- Foster, co-ordinate, and provide added value to the activities of BME member organisations and individuals, and collaborate where appropriate with national and international organisations, particularly the International Federation for Medical and Biological Engineering
- Serve and promote BME education, training, and accreditation programmes
- Serve and promote BME research and development
- Promote collaboration between BME Spanish research groups
- Establish recommendations for the appropriate general responsibilities, organisational relationships, and roles of those engaged in the field of BME
- Establish and maintain a liaison with the Spanish government and its agencies
- Promote public awareness of BME in Spain
- Improve relations and co-operation of SEIB members in Europe with other BME education and research institutions
- Encourage and assist in the formation of BME research groups and centres in Spain

To achieve these goals, SEIB is performing the following activities:

1. Organise the Annual Spanish Conference on Biomedical Engineering (CASEIB). This conference is the Spanish forum for scientific exchange of knowledge, expert meetings, and industrial promotion in the BME area. CASEIB is endorsed by IFMBE and generates the Annual CASEIB proceedings.
2. Prepare reports on the status and impact of BME upon the Spanish society
3. Apply for, participate in, and manage projects related to the promotion of BME
4. Collaborate with Spanish authorities, agencies, and organisations for the promotion of BME and for the allocation of funding and professional opportunities for the SEIB members and all the Spanish BME community
5. Endorse and support Spanish BME meetings
6. Contribute to the harmonisation and development of BME curricula

7. Identify and disseminate examples of good practice in BME
8. Act as a source of expert opinion on BME education, training, accreditation and application

Madrid, December 17, 2008
Enrique J. Gómez

Secretary-General SEIB
http://www.seib.uv.es/

Sociedad Española de Ingeniería Biomédica (SEIB)

La Sociedad Española de Ingeniería Biomédica- SEIB fue fundada en 1981 y es la asociación española en esta disciplina con un marco de actuación que se circunscribe a toda España.

Los objetivos de la SEIB son:

- Promover la Ingeniería Biomédica (IB) en España
- Promocionar, coordinar y proporcionar un valor añadido a las actividades de sus socios, tanto organizaciones como individuos, siempre que sea necesario y en un marco nacional e internacional, especialmente en la Federación Internacional de Ingeniería Médica y Biológica (IFMBE)
- Fomentar la educación, formación y programas académicos en IB
- Ayudar a la investigación y el desarrollo de actividades en IB
- Promover la colaboración entre grupos de investigación en IB
- Establecer recomendaciones para todos aquellos involucrados en la IB
- Establecer y mantener relaciones con el gobierno español y sus correspondientes agencias relacionadas con los ámbitos de actuación de la sociedad
- Promover el conocimiento en la sociedad española de la IB
- Mejorar las relaciones y cooperación de sus miembros en Europa con otras instituciones de educación e investigación
- Ayudar a la formación de nuevos grupos de investigación en IB en España

La consecución de estos objetivos se fundamenta en las siguientes actividades de la SEIB:

- La organización del Congreso Anual de la Sociedad Española de Ingeniería Biomédica (CASEIB). Este congreso es el foro científico en IB en España para el intercambio científico de conocimiento, experiencias y promoción de la I+D+i en esta disciplina. CASEIB está soportado y auspiciado por la IFMBE y publica un libro oficial de actas de los trabajos científicos
- Preparar informes del estado e impacto de la IB en la sociedad española
- Promover la participación en proyectos de IB
- Colaborar con las instituciones españolas para la consecución de fondos de investigación y educación para actividades en IB
- Auspiciar y apoyar reuniones en IB
- Contribuir a la harmonización de programas académicos en IB
- Identificar y diseminar buenas prácticas en IB

Actuar como expertos en educación, investigación y acreditación

Madrid, 17 de diciembre, 2008
Enrique J. Gómez

Secretario General, SEIB
http://www.seib.uv.es/

The Swedish Society for Medical Engineering and Physics

The Swedish Society for Medical Engineering and Physics (MTF) is working in the fields of medical physics, biomedical engineering and biophysics. The society was first founded in 1956 and it was called at that time "The Swedish Society for Medical Physics and Medical Engineering". The society was also affiliated to the Royal Swedish Society of Medicine in 1956 and hence, MTF is one of their oldest sections. In the beginning, the members of MTF had technological, medical or scientific background and their main interest was in the fields of research and higher education. In 1971 a subdivision, "The Division of Engineering in Health Care" was founded to meet the demands, not only from the persons who were working with clinical engineering at the hospitals and engineers from industry manufacturing biomedical equipment, but also from the medical doctors working with biomedical equipment in their every day life. Since the activities of these two groups within the main society served the same purpose the annual meetings 1981 decided, that the groups should join together and establish a mutual society named The Swedish Society of Medical Engineering and Medical Physics. At the annual meeting 2009 the name was changed to The Swedish Society for Medical Engineering and Physics.

Clinical Engineering and Health Care Technology Assessment Division, participants in Umea

MTF has about 1000 members and five sister societies, Denmark, Norway, Finland, Iceland and Estonia. The association arranged an annually recurring event called "The Biomedical Engineering Days". This event attracts about 350-500 participants from health care, research, industry and a large number of exhibitors. In addition, MTF participate at the "Sweden X-ray-week" and contribute with both own sessions as well as in conjunction with other compounds.

One of the association's main goal is to increase understanding and knowledge of security and safety in use of medical devices. MTF arranges several post graduate courses during each term. Some of these courses are recurrent like "Medicine for Engineers" and some are arranged only once, often required by new laws and regulations or new technologies. Most of the courses are organised by the members at their hospitals or at the local universities. MTF's program deals with actual issues in the clinical engineering for example Quality Assurance (QA), Quality Audit and accreditation of hospital laboratories. A topic of today is how Biomedical Engineering and Information Technology will work together on the issue of ensuring patient safety. Furthermore, MTF support the Swedish Medical Association with advice on referrals concerning medical and biomedical technology.

In 1993, the Annual Meeting approved a proposal according to which members have possibility to apply for Certification in Clinical or Biomedical Engineering. Each application to become a Certified Clinical or Biomedical Engineer is assessed by an Accreditation Panel. This panel place considerable emphasis on checking that the candidate has not only the formal education, but also the competence to be certified.

Clinical Engineering and Health Care Technology Assessment
Division teleconference, participants in Brazil

The engagement of MTF in international co-operation is formalised through the International Federation for Medical and Biological Engineering. However, several research institutes have contacts with other organisations and institutes. In 1967 the society arranged a World Congress in Biomedical Engineering in Stockholm and applied for 1997 World Congress. But in Kyoto meeting 1991 the application was withdrawn and MTF did offer full co-operation for Nice'97. MTF have hosted the Nordic Baltic Conference in Biomedical Engineering (NBC) three times; 1981, 1993 and 2005. These meetings have the status of IFMBE regional meeting. MTF also support other Nordic-Baltic countries arrangements of the Nordic Baltic Conference on Biomedical Engineering and Medical Physics. MTF intends to always have members who are active in the IFMBE divisions and committees.

Svensk förening för medicinsk teknik och fysik (MTF)

Svensk förening för medicinsk teknik och fysik (MTF) arbetar inom områdena medicinsk fysik, medicinsk teknik och biofysik. Föreningen bildades 1956 under namnet ”Svensk förening för Medicinsk Fysik och Teknik”. Medlemmarna hade medicinsk, teknisk och naturvetenskaplig utbildning. Föreningen blev samma år medlem i Svenska Läkaresällskapets sektion för Medicinsk Teknik och Fysik och är därmed en av de äldsta sektionerna inom sällskapet. 1971 tillkom ”Avdelningen för Teknik i Sjukvård” som en underavdelning. Medlemmarna var läkare, ingenjörer och tekniker verksamma med medicinsk teknik inom sjukvården och industrin. 1981 beslutades att grupperna skulle gå samman och bilda en gemensam intresseförening under namnet Svensk föreningen för medicinsk teknik och fysik.

MTF har ca 1000 medlemmar och fem systerorganisationer i Danmark, Norge, Finland, Island och Estland. Ett av föreningens största arrangemang är den årligen återkommande konferensen Medicinteknikdagarna. Evenemang lockar cirka 350-500 deltagare från sjukvård, forskning, industri samt även ett stort antal utställare. Dessutom deltar MTF i den Svenska röntgenveckan där MTF bidrar med egna sessioner men även i samverkan med andra föreningar.

NBC 2005 Opening Ceremony, Umea, Sweden

En av föreningens främsta mål är att öka förståelsen och kunskapen om säkerhet och trygghet vid användning av medicintekniska produkter. MTF anordnar flera kurser inom ämnet under varje termin. Några av dessa kurser är återkommande som "Medicin för ingenjörer" och några arrangeras bara en gång, ofta pådrivna av nya regulatoriska krav och förordningar. De flesta kurserna anordnas av medlemmar på deras sjukhus eller vid lokala universitet. MTFs program behandlar aktuella frågor i den kliniska verkstadsindustrin, t.ex. kvalitetsgranskning och ackreditering av sjukhuslaboratorier. Ett aktuellt ämne i dag är hur medicinsk teknik och informationsteknologi ska samarbeta inom sjukvården för att garantera patientsäkerheten. Dessutom stödjer MTF Svenska Läkarsällskapet med utlåtande på remisser om ärendet avser eller berör medicinsk teknik.

MTF utför sedan 1993 en frivillig certifiering av ingenjörer och civilingenjörer som arbetar med medicinsk teknik. Varje ansökan om att bli certifierad bedöms av en certifieringskommitté. Denna kommitté lägger stor vikt vid att den sökande inte bara har formell utbildning enligt givna krav, utan också den yrkeserfarenhet som krävs.

Föreningens internationella samarbeten sker via The International Federation for Medical and Biological Engineering (IFMBE), men även andra samarbeten existerar. 1967 var MTF värd för världskonferensen i medicinsk teknik i Stockholm. Föreningen ansökte även om värdskapet för 1997 års världskonferens. MTF har varit värd för Nordisk-Baltiska konferensen i medicinsk teknik (NBC) tre gånger, 1981, 1993 och 2005. Dessa möten har status som ett regionalt möte inom IFMBE. MTF stödjer även andra nordisk-baltiska länders arrangemang av NBC. MTF har för avsikt att alltid ha aktiva medlemmar som arbetar inom IFMBEs divisioner och kommittéer.

NBC 2005 venue, Umea, Sweden

The Swiss Society for Biomedical Engineering (SSBE)

The Swiss Society for Biomedical Engineering (SSBE) was founded in 1970, and is today well established at the Swiss universities and companies. Biomedical engineering itself is a critically important discipline in Swiss industry and education and, therefore, a central body to represent the interests of this sector within Switzerland is particularly relevant. One of the goals of the Society is to build a network for education, research, manufacturing, and politics within the background of biomedical engineering (BME). With the organization of regular scientific meetings, all these fields are brought together. Annual scientific prizes stimulate younger members to play an active role in the Society. In the interest of promoting international collaboration, the Society has participated in the organization of joint meetings with the Austrian and German Societies for Biomedical Engineering.

Two long terms plans of the Society are a stronger integration of small and large companies in the SSBE, and to become an important partner with respect to education in the different universities and universities of applied sciences (UAS), as well as the two Swiss Federal Institutes of Technology in Lausanne and Zürich.

With this in mind, the SSBE aims at establishing contacts, exchanges of information and experience, as well as cooperation between industrial and scientific people who want to apply physical and technical concepts to biology and medicine.

The SSBE currently has 272 individual.

Taiwanese Society of Biomedical Engineering

Development and Progress of Biomedical Engineering in Taiwan

The biomedical engineering (BME) education in Taiwan was provided for more than 36 years since the first undergraduate program established by Dr. Paul Han at Chung Yuan Christian University in 1972. To date, there are 18 universities offering BME programs with BS to Ph. D. degrees. Taiwanese Society of Biomedical Engineering (TSBME) was founded also by Dr. Paul Han on December 25, 1980, and was affiliated to IFMBE in July, 1985. The TSBME aims to promote biomedical engineering technologies and academic activities, and to activate BME industries for improving quality of healthcare in Taiwan.

With nearly three decades of efforts, TSBME has established thorough organization and operation procedures to provide services to our members and to achieve our aims. A brief summary for the development our society is as follows:

1. Member is categorized as group, life, individual, and student members. To date, active members are more than 600.

2. The elected president and board members serve for a two-year term. Our society organizes six committees in charge with member, BME promotion, journal publication, academy, certification programs, finance, and international affairs.

3. Publications. TSBME has an official journal, Journal Medical and Biological Engineering (JMBE), which currently is with Science Citation Index Expanded (SCIE). Textbooks: Introduction to Biomedical Engineering (in Chinese, ISBN : 9789866507038); Biomedical Engineering and Applications.

4. Conferences. Our society holds a regular Annual Biomedical Engineering Technology Conference taken place on the mid December of each year with the numbers of attendee increased to 1300 in 2009, and several promotion and training symposiums in clinical engineering, medical equipments and safety, and other specific academic topics.

5. Certification Program. Our society presently initiated three certification programs, including clinical engineer, biomedical engineer, and medical equipment technician in 2004. The certification programs aims to consolidate better services, quality, safety issues, and developments of medical devices of healthcare in hospitals and industrials.

6. Corresponding to the increase popularity of BME education and industrial developments, TSBME aims to provide further comprehensive services to our members, to promote publicity and journal, to collaborate and interact

with international BME institutes and societies more frequently, and to hold international conferences (upcoming conferences: WACBE World Congress on Bioengineering 2011; 9th Asia Pacific Conference on Medical and Biological Engineering in 2014).

中華民國生物醫學工程學會

臺灣生物醫學工程的推動與進展

臺灣在發展生物醫學工程方面，韓偉博士(1928~1984)以其卓越的智慧與服務熱忱，於1972年在臺灣中原大學首先設立生物醫學工程學系，是創始這塊園地的先驅者。自韓博士所設立之第一所生物醫學工程系至今，臺灣生物醫學工程教育的演進與發展已超過38年，現在共有18所大學院校已設立生物醫學工程學系與研究所，提供大學、碩士、以及博士學位。成立”中華民國生物醫學工程學會”（TSBME）也是由韓偉博士，結合一群熱心於醫學與工程整合之有志之士，共同於1980年12月25日創設。之後，並於1985年7月加入國際生物醫學工程聯盟（IFMBE）。生物醫學工程學會創立的目的，是在促進生物醫學工程技術、學術活動、及產業等項目的發展，以提升臺灣的醫療照護品質。

歷經30年的努力，生物醫學工程學會已經建立完整的學會組織架構及運作程序，提供本學會會員相關服務，除具體達成本會之創設目的外，更逐步提升本會之服務品質與內容。本學會主要的發展項目簡述如下：

1. 會員：分為團體會員、永久會員、個人會員、及學生會員。至今保持超過800位常態會員數。
2. 組織：理事長及理監事委員經全體會員選舉產生，任期為兩年。組織架構共分會員、證照甄審、學術、財物、學刊編輯、醫學工程推展、及國際合作等七個工作委員會。
3. 刊物出版：本學會之學刊”Journal Medical and Biological Engineering (JMBE)”從2009年7月起，正式被收錄於Science Citation Index（ＳＣＩ）。 此外，本學會亦編輯出版”生物醫學工程導論” 及 ”醫學工程與應用” 等大學用教科書（中文）。
4. 會議舉辦：本學會固定於每年12月舉辦”生物醫學工程科技研討會” ，與會人數持續增加，2009年參與者人數達到1300人。此外，每年也主協辦臨床工程、醫療儀器和安全、及相關領域的學術研討會，提供進修教育學分。
5. 證照認證服務：本學會於2004年起提供臨床工程師、醫療設備技師、及生物醫學工程師等證照認證服務，此項服務可進一步強化醫療單位更好的服務品質、安全維護問題、及促進醫療器材產業的發展。

因應生物醫學工程教育與產業發展日益受到重視的趨勢，本學會未來的工作將持續提供會員更完整的新知傳播服務、持續促進本學會的社會肯定度與互動關係、提升JMBE的學術影響力、保持與國際生物醫學工程機構建立經常性的互動交流、並爭取舉辦國際學術研討會。本學會於近期將舉辦之國際性活動包括第五屆世界華人醫學工程研討會(2011)，以及第九屆亞太區醫學工程研討會(2014)。

The Ukrainian Biomedical Association “Institute of Medical Engineering and Clinics”

The Ukrainian Biomedical Association “Institute of Medical Engineering and Clinics” was founded in Kharkiv, Ukraine in 1996. The Association joined the International Federation for Medical and Biological Engineering (IFMBE) in 1997. The "Institute of Medical Engineering and Clinics" is a public, non-profit organization whose members are specialists involved in research, educational and practical activities in the field of biomedical science, technologies, engineering, and their practical application. The objective of the Association is to realize the creative potential of its members in the interest of solving actual problems in medical science, through technology and engineering and to find technologies for practical applications in health. Among other, the tasks of the Association include:

- Developing measures for improving knowledge of individuals dealing with medical science, technologies, engineering, and practical application
- Developing priority lines of scientific research in the field of medical science, technologies, engineering, and practical application
- Accomplishing scientific, practical and financial assistance to members of the Association, to young scientists and specialists
- Assisting in improvement of the educational system and improving of professional skills of specialists;
- Developing cooperation between the state governing bodies and national scientific societies and associations.

The Association is based on a volunteer principle, which is expected to be developed further. Currently, the Association has 32 members.

Українська асоціація медичної і біологічної інженерії "Інститут інженерних і клінічних проблем медицини"

Українська асоціація медичної і біологічної інженерії „Інститут інженерних і клінічних проблем медицини" була створена у 1996 році в Харкові. У 1997 році Асоціація увійшла до Міжнародної федерації медичної і біологічної інженерії і набула статуту Національної. Асоціація є некомерційною, суспільною організацією, об'єднуючою на добровільних засадах фахівців, що займаються науково-дослідною, педагогічною і практичною діяльністю в області біомедичної науки, техніки, технологій і їх практичних застосувань. Метою створення Асоціації є реалізація творчого потенціалу його членів на користь вирішення актуальних проблем медичної науки, техніки і технологій в цілях їх практичних застосувань. Для досягнення вказаної мети Асоціація вирішує наступні завдання:

- здійснення заходів, направлених на підвищення знань осіб, практична і наукова діяльність яких пов'язана з медичною наукою, технікою, технологіями і їх практичними застосуваннями;
- здійснення наукової, практичної, правової і фінансової допомоги членам суспільства, молодим ученим і фахівцям, сприяння вдосконаленню системи освіти і підвищення професійної кваліфікації фахівців;
- розвиток співпраці з міжнародними і національними науковими суспільствами і асоціаціями.

З моменту свого заснування Асоціація функціонує виключно на волонтерських засадах, що ми намагаємося підтримувати і надалі. Зараз Асоціація налічує 32 члена.

Epilogue

Finally, all contributions for the Anniversary Book aiming to celebrate the 50th Anniversary of the International Federation for Medical and Biological Engineering have been collected and structured so that the chapters are ready for publishing. This book is meant to catch a moment in time in the development of medical and biological engineering, the state of art of biomedical engineering as it was in 2009, and also a glimpse into the future as seen from leaders in the field.

This book would never have been put together without the efforts of numerous distinguished colleagues, all volunteers, from Biomedical Engineering Societies from all over the world. The book editors sincerely thank all authors for their contributions.

When we started preparing the book, we had an idea of the kind of contributions we wanted to include. After the first set of contributions was collected, however, we realized that there should be more information on particular activities and persons, so we started collecting a second round of contributions. After collecting most of the contributions, and having the presentation of the texts from the biomedical engineering societies (bi-lingual in most cases), the hard times began and included editing the contributions, making the layout of different contributions alike, collecting appropriate photos with the text, reminding society representatives to send the logo of their society….

Therefore, it took us longer than we expected to prepare this book.

There are a number of colleagues I want to thank for their help in creating this book: Prof. Herbert Voigt, who developed the ideas on the presentation of the book, made the initial efforts to collect the contributions, and who always supported me in all efforts regarding the book. He also involved his Boston University proof-reading students, Timothy Mon and Aaron Silva. Prof. Paul Fagette of Temple University, a biomedical engineering historian, shared with us ideas on the how to best present IFMBE and the affiliated biomedical engineering societies and their activities. Dr. Christoph Baumann, the contact person for our publisher Springer, showed great patience and flexibility while we were working in order to complete the book. Prof. Alan Murray, who read the text and made a lot of suggestions to improve the text and the layout. I have been helped in preparing the book by the following young associates and students from University of Zagreb: Nikolina Hőlbling, Maja Sekulić, Dubravko Hendija and Milan Pleša.

Ratko Magjarević — Zagreb, July 2013

MIX
Papier aus verantwortungsvollen Quellen
Paper from responsible sources
FSC® C105338

If you have any concerns about our products,
you can contact us on
ProductSafety@springernature.com

In case Publisher is established outside the EU,
the EU authorized representative is:
Springer Nature Customer Service Center GmbH
Europaplatz 3, 69115 Heidelberg, Germany

Printed by Libri Plureos GmbH
in Hamburg, Germany